# Ferment

*Also by Tim Spector*

Identically Different

The Diet Myth

Spoon-Fed

Food for Life

The Food for Life Cookbook
*(with ZOE)*

# Ferment

*The Life-Changing Power of Microbes*

TIM SPECTOR

JONATHAN CAPE
LONDON

1 3 5 7 9 10 8 6 4 2

Jonathan Cape, an imprint of Vintage, is part of the
Penguin Random House group of companies

Vintage, Penguin Random House UK, One Embassy Gardens,
8 Viaduct Gardens, London SW11 7BW

penguin.co.uk/vintage
global.penguinrandomhouse.com

First published by Jonathan Cape in 2025



Typeset in 12/14.75pt Bembo Book MT Pro by Six Red Marbles UK, Thetford, Norfolk
Printed and bound in India by Manipal Technologies Limited

The authorised representative in the EEA is Penguin Random House Ireland,
Morrison Chambers, 32 Nassau Street, Dublin D02 YH68

A CIP catalogue record for this book is available from the British Library

HB ISBN 9781787334656
TPB ISBN 9781787336131

Penguin Random House is committed to a sustainable future
for our business, our readers and our planet. This book is made
from Forest Stewardship Council® certified paper.

To my long-suffering wife, Veronique –
sorry about the state of the fridge!

## Contents

# Introduction

Of all the ways to prepare food, fermenting is surely the most mysterious, miraculous and misunderstood, yet humans have been finding ways to ferment plants, dairy and meat for thousands of years as a means of preservation and enhancing flavour. Fermentation simply means the chemical transformation of any food or drink, with the help of yeast, bacteria or other microbes, often producing bubbles or heat. Whole foods already contain hundreds of compounds; however, once bacteria or fungi have worked their magic, these compounds multiply and foods become vastly more complex. As these fermenting microbes feed on the food or drink, they produce hundreds of new and unique compounds. Wine is infinitely more complex in flavours and chemicals than grape juice, as is cheese compared to milk. We now know that this ancient process of alchemy not only transforms the flavour of the food, making it more complex, varied and delicious, but it also brings a multitude of additional health benefits.

In 2024, one of the largest ever trials of a new supplement was performed, with nearly 10,000 UK volunteers. For three weeks they were asked to take three doses of the supplement daily and monitor changes to their health. The results were amazing. Nearly half of participants (47 per cent) saw improvements in mood, 55 per cent reported more energy, 52 per cent less hunger, and 42 per cent a decrease in bloating. In fact, this trial was organised by my team and ZOE (the science and nutrition company I co-founded), and it wasn't a new commercial supplement we were testing, but ordinary shop-bought fermented foods, such as sauerkraut, kefir, yogurt and kimchi. If these results had been for a new vitamin supplement it would be a blockbuster. That all these benefits could come simply from adding such humble ingredients to your diet is even more astonishing. In this book, I want to demystify fermented foods and explore why it is that they bring so many benefits. In fact, I hope to convince you to throw away your vitamin supplements and instead try thinking of fermented food as a vastly more nutritious and tasty supplement, with many more proven benefits for your health.

I now get asked about fermented foods more than nearly any other topic. It is these potential health properties and their effect on gut health that has ignited people's interest. Although I have mentioned fermented foods in previous books, the science is evolving so fast that there is much more to say. For instance, I was recently blown away by research showing that even some dead microbes in our foods can still be beneficial. This means that some beers with dregs of dead yeast in them could have some health benefits, which might partly help to compensate for the negative impacts of alcohol. These kinds of breakthroughs in our understanding of the benefits of fermented foods have not yet been given the prominence and space they deserve.

Perhaps one of the reasons that these benefits aren't better known is that many of us are fearful of fermented foods. We've been conditioned to associate them with strange tastes and smells, dangerous bacteria, and slimy, mysterious microscopic creatures. There is much confusion and misunderstanding about how to consume them too. People often ask me if they work at all, given that stomach acid destroys the live microbes. Aren't probiotic supplements more effective? Isn't all that saturated dairy fat in kefir and yogurt bad for you? Don't kombuchas rot your teeth, cause acid reflux or food poisoning? Are they not dangerous if you have cancer or autoimmune disease?

Before 2010, when I first started studying gut microbes, I have to confess I was ignorant about fermented foods. Like most people I had seen over-hyped adverts for commercial yogurts that I assumed had no real health benefits. I had no idea that my Stilton contained health-giving fungi in its blue stripes, or that sauerkraut might contain live microbes. I had not heard of kefir or kombucha, and kimchi was for me just a spicy pickle to be avoided. I enjoyed Japanese food but had no clue that microbial ingenuity was behind the complexity of soy sauce, tofu and miso soup. Being a medical doctor trained in the 1970s I did know quite a lot about alcohol, professionally and personally.

While I knew microbes were involved in the fermentation process to change sugar into alcohol, and could also turn forgotten red wine into vinegar, I didn't realise that it was fermentation that made my humble cup of coffee so drinkable and yet beneficial for long-term health, and the reason my sourdough bread was both tastier and healthier than supermarket loaves. As I delved deeper, more surprises followed: Marmite and Tabasco sauce, it turned out, were also fermented products.

I started to experiment with fermented foods in my own diet and soon became converted to the cause. I began making my own kombucha with my trusty SCOBY (symbiotic culture of bacteria and yeast), affectionately known in my house as Blob, which I nurtured from a tiny baby blob I had found in the dregs of another bottle. Before long, I found myself adding fermented vegetables to almost every meal and enjoying my sour morning kefir (fermented milk) with nuts and berries more than any other breakfast. I made my own version of kimchi (known in my household as Timchi) from fridge leftovers and found I loved the taste, as well as its benefits for avoiding food waste. I was glad to discover that red wine and maybe even artisanal ciders (in moderation) have a beneficial effect on our heart and our microbes. I now find that I have to be doing some regular fermenting to feel happy as well as healthy.

Doctors like me used to be taught that all microbes were dangerous and our job was to wipe them out, whether inside or outside the body, with antibiotics and antifungals, and that our food should effectively be sterile in order to be safe. I know from emails I receive that many doctors are still worried about some patients eating even yogurt. We have forgotten that for millennia, before the widespread use of fridges, our ancestors purposefully *added* natural live microbes to make their food safe to eat. They understood what we have since forgotten: that microbes are generally our friends. If we help them along by manipulating the environmental conditions, they will rapidly colonise a cup of milk or some vegetables in salty water and produce chemicals that change the acidity or, when put into sugar, produce acetic acid or alcohol, which allows only friendly microbes to grow. This stops the food going rotten, keeps more harmful species away and adds greater complexity of flavour. It also – we now know – brings extraordinary benefits for our health, benefits that we scientists are only just beginning to understand fully.

For many of us, at least in the UK, fermenting has become a bit of a niche pursuit. My first encounter with fermentation was aged about sixteen when my parents bought me a home brewing kit, probably thinking it was safer for me to get slightly tipsy with friends at home than going to the sort of pubs where they didn't ask your age. It turned out that I might have been safer there, as what my brews lacked in taste and clarity, they made up for in power, leading to a very special type of hangover. My father (a medical scientist) brought home an old 40-litre storage vessel from his lab and, after some disastrous efforts and odd microbial overgrowths, I had a brew that was drinkable, albeit for an undiscriminating teenager and his friends. I still remember the distinctive tastes of the malt in a tin and the yeasty ferments at the bottom of the container, which accompanied me to my first year of medical school, before it was discarded for other social pursuits and better ale. I would never have guessed that decades later I would be writing a book on the subject of fermenting, and that I would be evangelical about its benefits for our health.

Now, in the 2020s, there is growing curiosity about ferments, but the world of fermentation can seem bewildering too. Some of this is down to the brilliant but deceptive marketing of products such as ultra-processed yogurts with a dozen added chemicals and fake fruits, which claim to be healthy because of a few strains of microbes added at the last minute before packing. Many food companies try to entice us by using words like microbes and fermentation. We often do not know if by the time we buy a 'fermented' food the microbes are dead or alive, as there are pitifully few food labelling rules to help ensure the consumer is informed.

When it comes to experimenting with fermenting ourselves, many of us don't know where to start. Perhaps you'd like to try fermenting yourself but are afraid of kitchen explosions. Perhaps you tried making sauerkraut once, but didn't like the taste or were worried it would make you ill. My hope with this book is to persuade you to try fermenting at least one food or drink yourself, even if it's just sticking some garlic cloves in a pot of honey. I will show you how ferments work in your own body and can improve your health, and why we should all aim to consume at least one fermented food every day. And even if you never plan to experiment yourself, I will help you spot the best (and worst) products in shops, and at the very least increase your admiration for the humble microbes – inside our bodies and in our food – that do all the work.

I often talk about the importance of the four Ks: Kefir, Kimchi, Kombucha and Kraut, but in researching this book I have come across hundreds of ferments that were new to me and discovered many new tips and recipes. At the end of the book I've shared the best and simplest recipes that you can try at home, with minimal equipment and simple ingredients, ranging from making your own style of kimchi to honey-fermented garlic. I want you to learn from my trials and errors, to encourage you to roll up your sleeves and try fermenting to improve your own health too.

My main focus is on foods that contain live microbes when they are served, rather than those that get killed by baking or roasting, boiling or in alcohol. But, as the new science now shows us that certain microbes can still have health benefits beyond their grave, I will also tell you more about these 'dead' or 'not-so-live' microbe products that could still provide life after death. The chances are that these are foods that you already enjoy, whether it's coffee, tea, sourdough bread, wine or beer.

This book is intended to titillate your imagination, taste buds and gut microbes and expand your understanding of what our ancestors called cold cooking. By understanding more about this cooking that occurs outside your body in jars you will learn much more about the cold cooking that occurs every day inside your body and how important it is for your life and soul.

But first, to better understand what makes a fermented food, let's begin with something that is very much alive, teeming with different life forms . . .

# PART ONE
# What is Fermentation?

## Voyage of the microbes – from field to fermenting jar

Picture a field in the countryside at sunset. The late summer sun is hazy and paints orange hues on the outermost leaves of the humble purple cabbage. It sits proudly above the soil that has been its home for the last six months. The soil and its microbes, with the infinite web of mycelium, have supported its growth from tiny seed into a complex, multi-leaf, colourful structure. Its intricate food matrix holds everything together in a neatly packed pattern. The deeply coloured purple leaves hold hundreds of polyphenols (defence chemicals), which have protected the cabbage from any damage resulting from changes in temperature and availability of water, and which will go on to provide additional fuel for the microbes living in the gut of anyone lucky enough to eat it.

This red cabbage is ready to be harvested. Whether it was grown with organic or small-scale farming methods with minimal use of pesticides and fertilisers, or sprayed with both to encourage a quick harvest, tens of millions of microbes will have colonised the folds of the leaves and its crevices. The cabbage is picked, washed and packed up to be transported to supermarkets or farmers' markets, in plastic or wooden crates, almost always refrigerated for freshness.

By the time the red cabbage meets its final owner – you or I – the outer leaves will have come into contact with washing agents, hands, plastic, transport, more hands and maybe some supermarket shelves. All these processes are likely to affect the microbial diversity of the outer sections, but plenty of microbes remain. Now it is on your chopping board and you're about to transform this wonderful piece of natural architecture into something even more extraordinary.

First, you remove the outer leaves of the cabbage (keep them for later) and give everything a good rinse under a running tap. Next, you chop the cabbage into thin slices, admiring the beautiful patterns that lie inside and knowing that there are millions of tiny microbes in those swirls, living off the sugars in the cells of the cabbage. Microbes will be plentiful in any cabbage you buy (whether organic or not) and they are specially adapted to that plant. They come in many types – those most likely to be successful are the group known as lactic acid producing bacteria (LAB for short) and are called *Lactobacillus*, *Leuconostoc* and *Weissella* ; other less well trained (adapted) ones are called *Klebsiella*, *E. coli*, *Pseudomonas*, *Bacillus* and *Staphylococcus*. You will usually find some yeasts called *Saccharomyces*, which also like beer, as well as *Candida*, which likes dark sweaty bits of our bodies. You will also find some moulds, like *Aspergillus* and *Penicillium*, and finally many tiny viruses called phages that are highly specialised killers of the lactic acid bacteria. Think of this motley crew as competing teams in a fermentation game show, where teams of microbes will compete for survival.

## The first fermentation

Once chopped, you simply place the cabbage into a clean bowl with a generous pinch of sea salt (around 2 per cent of the weight of the cabbage) and start to massage the salt in with clean dry hands. You squeeze and massage your cabbage well, drawing out as much juice as possible. The salt causes the plant cells to lose their structure as the internal fluid leaks out under a process called osmosis. After a few minutes you should have a good amount of liquid at the bottom of your bowl, which will be full of natural sugar from the cabbage – an exciting fast-food treat for the microbes to help them grow and multiply. You transfer the floppy cabbage pieces with all their juices to a clean jar, packing it down with your fist, making sure the liquid covers the top of the cabbage. You might top the jar up with a little salty water if needed, add some herbs or spices such as caraway seeds or fennel, and any other shredded veg you want to use. Just a few extra slices of carrot gives the microbes more tasty carbohydrate options to eat and speeds up the fermentation process. You then use the outer leaves of the cabbage to push down the mixture as hard as you can to compress it – this acts as a natural lid, keeping the air out.

We are surrounded by microbes. Earlier that day, if you were gardening or stroking your pet, some of the soil or pet microbes will have made your hands their home, even if you washed your hands. Your kitchen window is open, and the breeze carries in some extra microbes including natural yeast. These microbes too will now join the game show, competing with your tough cabbage microbes in a brutal survival contest to see which will give your fermented cabbage its unique composition. The first challenge is to see which microbe teams cope best without oxygen. Friendlier anaerobic bugs that can live without much oxygen are more likely to survive, while the oxygen-loving microbes, which include a few nastier characters (pathogens) that can cause health problems, will likely be killed off by the salt and the lack of oxygen in your fermenting jar. At the same time the salt, as well as punishing some microbes, offers survivors a reward by sucking out and releasing the sugars normally stored in the cabbage cells.

Now you pop your jar shut, reducing the oxygen inside, and know that many microbe competitors have now been eliminated, leaving more food for the survivors. All you do now is place your mixture into a cool, dark place and wait for the magic to happen. This process of 'cold cooking' isn't really magic but simply an elimination contest of chemical-producing microbes. After just twenty-four hours, you'll start to see bubbles forming at the top of the liquid – a clear sign that tiny microbes have reproduced and started feasting on the cabbage. Its fibres and polyphenols, as well as the sugary core, are a banquet for these bugs, and their appetite is insatiable. These tiny bubbles of carbon dioxide produced by specialised microbes (both yeast and a few bacteria) rise to the top and may need to be let out every few days to avoid too much pressure building up into a slow dribble of liquid or, worse, a ferment explosion. This is what's affectionately referred to as 'burping' your ferments.

The survival contest isn't quite over for the microbes, and there is still plenty of competition to get to the final prize. Some rivals have quite enjoyed the lack of oxygen and the salty environment and, without the competition from their oxygen-loving colleagues, they have multiplied. But they are in for a shock. When they are without oxygen, other microbes start producing weak acids, such as lactic acid and acetic acid, which change the mixture into an acidic environment with pH levels below 4.5, which few species can handle. After about three days, only the specialist bacteria and yeasts are left. The aggressive microbes that can cause trouble, such as *E. coli* and *Staphylococcus*, have been eliminated, making the now fermented cabbage – or sauerkraut – safe to taste-test.

Your purple cabbage has now turned a sunset pink and is softer in texture and tangier in flavour. The hundreds of different chemicals and fibres from the original plant have fuelled the survivor microbes to multiply furiously, producing even more acid and carbon dioxide bubbles. The winning microbes now call your jar home, creating hundreds of new chemicals, or metabolites, as a by-product of their demolition of the cabbage. Every microbe that breaks down fibre and uses polyphenols creates a so-called 'postbiotic' chemical, which never existed before this process began. Like a magician popping a bunny out of a hat, this is microbial wizardry at its finest and you have a front row seat. And this exact same process of fermentation, breaking down plants to make brand new chemicals, happens inside each and every one of us.

It's time to taste your tangy sauerkraut after patiently waiting for it to reach just the right amount of crunch, sourness and complexity for you. I eat sauerkraut after between five and ten days depending on colour (red cabbage takes longer than white), plant combinations and room temperature, which all impact the speed of fermentation. This wonderful cocktail of microbes and metabolites, fibres and polyphenols then enters your internal gut universe to face further struggles for survival in the next round of the fermentation game show. Waiting for them are hundreds of trillions of microbes (mainly bacteria and viruses, but also fungi and parasites in our large intestine, and before them, just a few billion microbes in the lining of our small intestine). Each mouthful of fermented food not only adds newcomers to join our resident gut microbiome, it also helps feed them with a delicious variety of fibres and polyphenols that have already been partly transformed or pre-digested by fermentation in the sauerkraut jar.

The difference between eating raw sliced red cabbage in a salad and fermented sauerkraut is huge – and this is all down to the power of the oldest form of cooking: fermentation. Let's look at what happens to the kraut and its special team of acid-loving, oxygen-hating survivors as we take a bite, and they are once again forced to adapt to a new environment in the finals of the competition.

## The second fermentation

In our mouths, mixing with our saliva, our competitors find themselves in a neutral, non-acidic environment. This isn't a problem as they are only there a few seconds before passing on to the next challenge: the stomach. Here they are surrounded by gastric acid, which has a pH lower than 2 – much too low for comfort for the microbes. We used to assume that virtually all the billions of microbes (25 billion on a serving of sauerkraut) were instantly killed off by the stomach's acid, but this was based on old data and cell culturing methods that ignored 99 per cent of microbe species. Some will die off in a few minutes, especially the ones from the outer parts of the kraut, but others may be protected within a layer of leaves where the acid can't penetrate. Also, very small microbes, which were only recently discovered, may fly under stomach acid's radar. Called ultramicrobacteria (UMB) or ultra-small bacteria, they have a maximum size of 0.1 cubic micrometres ($\mu m^3$) and some are as small as 0.009 $\mu m^3$. To put this in perspective, a single *E. coli* bacteria could house 150 of these tiny lifeforms, and 150,000 would fit on the tip of a human hair. Aside from UMB, many standard-sized microbes can also suddenly shrink down as a temporary protection mechanism against changes in acidity or temperature, usually avoiding detection by even our current methods.

We know that a few selected bacteria love even these harsh conditions and live in our stomach permanently; for instance, a key natural resident is *Helicobacter pylori*, which can cause ulcers and can usually be eliminated by a combination of three antibiotics. But, interestingly, probiotic microbes targeting the stomach lining can also reduce the severity of these ulcers, when the acidity has dropped. The acidity of our stomachs varies a lot and people with stomach infections (gastritis) or taking antacid drugs like PPIs (proton pump inhibitors) have reduced acidity, allowing many more good and bad microbes to survive. This can occasionally be good if the microbes come from probiotics, like our kraut, but they are more generally bad for us if they are pathogens, which increase our risk of gut infections.

So despite the cull of all the weaker, more exposed microbes, many of our merry band still hang on to pieces of cabbage and continue undaunted down to the duodenum and small intestine. Here, life gets somewhat easier in the more hospitable surrounding liquid and slimy gut layers. Our contestants will pass a few resident microbes in the small intestine tucked into the many crevices and hiding places of this vast structure, which, at 7 metres long and 200 metres squared, would nearly cover a tennis court if stretched out. We don't yet know much about precisely what happens in the (badly named) small intestine, as it is very hard to access without cutting someone open. New smart capsules, which are designed to be swallowed, take a tiny biopsy as they enter this zone and get collected in the toilet the next day, but so far only a few of these expensive experiments have been conducted. So while we know that the main role of the small intestine is to absorb nutrients, we are still guessing what else goes on here. The current consensus is that the main interaction between the intestine's residents and the fermented foods is signalling and sensing. The microbial residents, when detecting the cabbage visitors, signal to their friends and networks that fibre and polyphenols are on their way down to the colon (large intestine), which contains vast hordes of microbes.

Important collaborators with the resident microbes are newly discovered 'neuropod' cells in the gut lining, which are specialised at sensing what food is being ingested by the microbes and what chemicals are being produced. This means that, via a vast relay system, they can rapidly alert the brain – in milliseconds – about the good and bad nutrients coming their way. These novel food sensors have a key role to play in our eating patterns and have changed the way we think about the relationship between our brain and gut. As well as having taste receptors and detecting tiny amounts of sugars, fats and proteins, specialised neuropod cells can detect microbes and their chemical metabolites such as short chain fatty acids (SCFA).

A new idea is emerging that the main effect of fermented foods, or at least the probiotic microbe component, is the way the microbes act on the small intestine to influence and talk to the immune system. Although we lack hard evidence, this makes sense as this is the only place on the journey that the gut environment is not hostile to our food microbes and, importantly, they are not outnumbered a million to one by resident microbes that would out-compete them. Some diseases of the small intestine can be helped or sometimes cured by probiotics, so we know external microbes can work in this environment. So it makes sense that the food microbes travelling on their cabbage raft stop off at this friendly port to rest and reproduce. This allows them to expand in numbers, producing greater quantities of helpful chemicals that interact with both the neuropod cells and the immune cells on the gut lining.

So, as our friendly kraut package makes its way to the large intestine, our body has plenty of advance warning and is primed.

After a brief respite, our intrepid travellers – with somewhat boosted numbers – are pushed through the ileal valve by peristaltic waves driven by the extensive nerve connections into the dark depths of the colon or large intestine (which is actually shorter than the small intestine). Despite the advance warning signals, our travellers are in for a shock. Suddenly they are surrounded by a mass of colon microbes. Imagine a group of twenty football fans trying to stick together as they leave the stadium after an away game, when all they can see are cheering opposition supporters. The microbes try to hang on to the fermented cabbage life raft for dear life. If they fall off they will surely die, as they are used to living on acidic salty cabbage and have not adapted to live in the alkaline environment of our gut, unlike the hardy locals.

A lucky few microbes with access to undigested cabbage manage to hang around hidden and unnoticed to reproduce and produce some healthy chemical by-products such as short chain fatty acids with anti-inflammatory effects. As in the small intestine, these are detected by the neuropod-sensing cells and the many immune cells in the gut lining and send calming signals to our immune system and our brain. We believe these long-travelled microbes send these healthy signals via chemicals that are both powerful and unique; in other words, these chemicals would not be created simply by having your resident microbes feast on boiled cabbage. This is the power of fermented food.

Soon after, the kraut life raft is completely devoured by the hungry resident microbes, who use the nutrients in the different fibres and are able to extract all the polyphenols for energy to reproduce and produce even more beneficial chemicals. After a couple of days of fighting against the million to one odds in our guts, nearly all our original kraut microbes and their descendants will have likely died off. Their dead bodies are unceremoniously discarded in our stool, along with the trillions of other short-lived resident microbes. About half our stools every day are made of tiny dead microbes, which gives you an idea of the sheer numbers involved as well as the fast pace of life in our guts.

Until recently we couldn't seriously examine more than a fraction of microbes in a normal stool sample. But now, with genetic sequencing, we can estimate how many of our lactic-acid producing microbes in our ferment make it all the way to the toilet bowl. Regular consumers of fermented foods, whether it is simple yogurt, kimchi or kraut, have much higher levels of fermenting microbes in stool samples than non-eaters. Crucially, these numbers are much greater than the amount of microbes entering the gut, showing they have multiplied inside our bodies.

The old assumption was that all probiotic microbes (whether in food or capsules) that reached our stomachs, died off before they could be effective and have any real health benefit. But we now know that many of them survive the stomach and small intestine before they reach the colon, and countless clinical studies now show they can improve health. Another odd but compelling reason the probiotic microbes may be helping us is the possibility that they have an afterlife. Even after dying on the journey, we now believe they can provide post-mortem health benefits through chemicals they produce or from proteins on their cell lining. So far these zombie microbes only appear to be helpful, not harmful. We will explore this wacky afterlife idea more later.

One of the reasons fermented foods are better for us than just eating the raw form is that they are, in essence, double fermented. The first ferment is in the jar, which you can witness unfold with its new smells, bubbles and changes in colour, texture and flavour; the second you have to imagine – in the darkness of your bowels. That first ferment means that your resident gut microbes can skip all the boring prep work and get on with cooking their ideal meal, giving them all the nutrients they need. A simple example of this is the way the lactose (the sugar present in milk) is predigested by external microbes when you make yogurt, cheese or kefir, so our own gut microbes have much smaller bites of lactose to deal with and so are therefore much more effective in splitting it into tiny sugars that are easily absorbed. This is why many people with milk intolerance or lacking the lactase gene mutation (80 per cent of the planet's population) can cope with eating fermented dairy but not drinking fresh milk.

It has been a long journey for the microbes, from the cabbage in the field, to the salt-loving kraut microbes in a jar, to the heroic martyrs in the intestine who deliver their chemicals and die off. I hope their voyage helps to show you their transformational power and the astonishing efforts they make to keep us healthy.

## Meet the fermenters

The life of microbes is fast and furious; they can pack a lifetime into a sixty-minute action movie in which they are born, live, eat, fight, reproduce, excrete chemicals and die. But their legacy lives on, as the influence of their dead body parts or fluids can still be felt months later. Many microbes can enter a state of suspended animation as either shrunken versions of themselves or as larger spores with a thick coat, protected from the harsh environment (acid, heat or cold) until conditions improve again. Others we know little about are so small they pass through filters and are called ultramicrobacteria.

As many as 50 per cent of bacterial species are able to form spores and enter a form of hibernation that can last years. This is a cunning defence mechanism to protect them from hostile environments. In rare cases, these spores can cause problems for us such as food poisoning from a microbe in reheated white rice (*Bacillus cereus*), which is difficult to kill off with heat or alcohol and comes back to life with gentle heat. Luckily spores are not normally something to worry about when fermenting though, and most LAB are not spore-forming.

The speed at which microbes proliferate, and so the speed at which fermentation occurs, depends on several factors including the acidity of the environment and, crucially, the temperature. Some microbes can reproduce and duplicate every thirty minutes inside the body, but fermenters are a bit slower. After a few hours to settle into their new surroundings, they can double every ninety minutes. This means that in twenty-four hours twelve to thirteen generations can be produced. So if you add 3 grams of kefir grains containing 300 million colonies to your milk (we quantify numbers of microbes in colony-forming units, or CFUs), this works out after twenty-four hours at about 2 trillion microbes and over 700 billion per millilitre, which is pretty impressive. The number of microbes may be lower for most commercial kefirs and yogurts where microbes are added at the later stages of production, just before chilling. In comparison, most probiotic supplements only have 1–10 billion CFU per capsule. Eventually the microbes die off as they run out of nutrients or space or they are stopped by the conditions. They often auto-destruct by producing too many by-products, like increasing alcohol content (around 15 per cent is tops) or overdoing the acidity.

Until very recently we knew virtually nothing about how many food microbes there are and what role they play. A 2024 study, which was the culmination of five years of work from an EU consortium, finally answered some big questions and tripled the number of food microbes we know about. It created a food microbe database by sequencing bacterial genes from 2,533 different foods and comparing them to the genes in gut samples of 30,000 people. These ranged from alcohol to dairy, kombuchas and kefirs, fermented (sausages and salamis) and non-fermented meats, seeds, roots, fermented and non-fermented fish, and fermented and non-fermented vegetables and fruits. They also looked at 358 subtypes of probiotic foods (from kimchi to sourdough) and included some foods that had never been studied properly before like fermented seeds, fish (e.g. Korean skate), Mexican pulque (fermented agave), African palm wine and different meats. They found over 10,000 distinct bacterial genomes that grouped into 1,036 species groups, the largest and most common of which were the Bacillota (formerly and confusingly known as Firmicutes). Around half of these microbes were known and half novel, even in well-studied food, such as dairy and many cheeses. These results show just how much we are discovering and have yet to learn.

Around 40 per cent of the bacteria detected were picky and found in only one food, whereas around a quarter were found in multiple foods. In general the fermented foods contained microbes that appeared in multiple other fermented foods; in non-fermented foods these species were often quite specific to the food. Examples include *Brochothrix* in meat, which makes it rapidly go rotten and smell, but an exception was raw milk, which contained low levels of the microbes you find in yogurt, several strains of *Lactobacillus* and

*Streptococcus thermophilus*. Many of these microbes have never been detected outside of food, suggesting that they have adapted to live in these specific foods.

The variety of different species was also tested. The study actually found more diversity of microbe species on regular unpasteurised food compared with fermented food; this is because the process of fermentation rapidly eliminates the non-adapted microbes. The highest diversity of microbes was seen in fish samples (which explains the unique rotting fish smells due to the many chemicals produced). Unknown or novel species were found in nearly all foods, with the highest recorded numbers in fermented tea (kombucha) and Mexican fermented agave (pulque). Only twenty-five microbes (twenty-three bacteria and two yeasts) determined the tastes of most food categories and were more important than geography. Strangely there was considerable overlap of microbes living in kombucha, coffee and fermented (pu-erh) teas.

While hardly any food microbes were found in human saliva samples, around half were found in stool samples, suggesting that they can survive in our intestines, at least for a short while. In infants, 56 per cent of the total gut microbe species are shared with those in food, with a key microbe *Bifidobacterium longum* (found in breast milk) dominating, but as the gut microbe population increases in diversity, this decreases rapidly to 8 per cent in older kids and down to an average of 3 per cent in adults. This partly reflects an increasingly diverse diet – not just dependent on milk – but also our increased interactions with the wider world and the people and animals that inhabit it. This 3 per cent figure emphasises that the trillions of microbes in your gut can come from many different non-food sources, including the air, soil, animals and other humans. It also shows that the microbes we do share with food punch above their weight in their health benefits. As we saw earlier, the fermenting microbes are also likely to have a much bigger impact on the small intestine where (if we could measure them well) they would likely form a larger percentage.

But whatever the composition of your gut, you can acquire novel or extra microbes by eating different foods, whether fermented or not. A great example is a specialist microbe found in algae (*Vibrio* EJY3). When you eat fish that feed on this algae, you also ingest these microbes, which colonise your gut. Then, through a process called horizontal gene transfer, these microbes share DNA with gut bacteria already dwelling in your colon. This genetic game of give and take means that genes coding for enzymes that digest the seaweed get passed around, allowing your existing gut microbes (and you) to access all the nutrients and health benefits of seaweed when you eat it. This is what happens to people living on the coast in Japan, as their gut microbiomes evolve to benefit from the abundant and cheap seaweed around them, which they eat raw. People living elsewhere who occasionally eat sushi are unlikely to have the right seaweed-digesting microbes in their guts to extract all the best nutrients from the tough polysaccharides in raw seaweed. Our guts, depending on where we live, are also full of another unlikely passenger, brewer's yeast (*Saccharomyces cerevisiae*), probably because so many foods we eat regularly contain it.

We found another interesting microbe that is easy to pick up when studying a 30,000-person subset of the ZOE database of over 250,000 people with our Italian colleagues. We published our findings in the prestigious journal *Nature Microbiology*. The microbe *Lawsonibacter* is found in coffee and is present in high levels in the guts of coffee drinkers but not in tea drinkers. What we discovered is that although newborn infants lack this microbe, just a kiss from a coffee-drinking parent can colonise the gut in low levels. The microbe then waits patiently in a corner of our guts in small amounts for the day when it can feast on fragments of coffee. People in countries that never drink coffee had very low amounts of *Lawsonibacter*; this means that even if you never eat or drink a particular food or beverage, if it is consumed by those close to you, some of its microbes can make their way to your gut. This is another reason diverse food cultures are a positive thing.

Many of the microbes living in or on our food are just there for the ride and may not be involved in fermenting or transforming the food, but others have evolved these special skills. There are hundreds of these potential fermenters, ranging across a wide spectrum from bacteria to fungi and viruses. These fermenting microbes play a crucial, invisible role in transforming our food and our health. Despite their obscure-sounding scientific names, we need to start seeing them as our friends. So let's take a moment to get to know some of these different types of microbes and how they work.

## LAB (lactic acid forming bacteria) and other bacteria

***Lactobacillus*** Lacto means milk in Latin and bacillus means a rod, so its name is a clue that this important group (with twenty-five subsets) of rod-shaped bacteria loves milk, feeding off the lactose sugar component. It's also one of the LAB that are key for fermented foods as they convert sugars to acids and keep the Ph low (high acidity), thereby preventing spoilage by other microbes. They prefer dark unoxygenated spaces but can tolerate air, so they are called aerotolerant anaerobes. They prefer warm temperatures and tend to go to sleep in your fridge. They are fussy eaters and only feed off certain types of sugars called hexoses, but luckily there are plenty of these sugars available to them. Well-known subtypes include *L. acidophilus* (acid loving), *L. delbrueckii, L. casei* and *L. paracasei* (cheese loving), *L. rhamnosus* and *L. gasseri* and they are present in many ferments, such as yogurt, kefir, sourdough, sauerkraut, tempeh (fermented soy) and kimchi. They are used as probiotics and often form part of the starter culture for fermented foods. (Starter cultures are a mix of microbes and medium, i.e. food that the microbes enjoy, like seeds or grains. These well-established microbial communities kick off the fermentation process.)

***Lactococcus*** have a similar role and profile to *Lactobacillus* but grow in chains rather than rods. They also produce lactic acid, only eat glucose products and are commonly used in dairy fermentation. Many have *lacto* in their name as a clue, but there are many others, such as *Pediococcus, Aerococcus* and *Weissella and Leuconostoc* that have similar acid-producing properties.

***Leuconostoc*** are chain-forming but are more adaptable and able to eat different foods, so they can live off a greater variety of plants, such as cabbage and milks. They also help make kombucha when they form part of the SCOBY colony (more about this later). They are useful to the process but can sometimes make your sourdough starter rather smelly, but the stench is generally short-lived – as the acid-producing bacteria grow in number, acid-sensitive *Leuconostoc* will be killed off.

***Streptococcus thermophilus*** is a commonly known lactic acid producer, often used to make commercial yogurt. It is well adapted for poor oxygen environments, where it can hang around waiting for its moment. It is really fussy about temperatures though, and its superpowers come to life at warm temperatures of 35–42°C. It works well in a buddy system with its mate *L. delbrueckii*. Together, they have a non-aggression pact and swap nutrients such as folic acid. Many microbes produce vitamins like folic acid or vitamin K and use these as energy to trade with their friendly neighbours, as well as making some of these vitamins available for their hosts (that's us).

What happens when an LAB such as *Lactobacillus* meets a glass of milk, for example when we make kefir or yogurt? As the bacteria floats around in the milk it will absorb (swallow) a small globule of the lactose sugar; once inside, it will digest it with an enzyme into its component sugars: glucose and galactose. It then uses the glucose and (sometimes the galactose) as an energy source to rapidly grow and reproduce and in doing so it releases lactic acid as a by-product. This causes a build-up of acid inside the cell, which it releases into the milk, slowly making it more acidic. This discourages any other less adapted microbes from feasting on the same milk.

***Acetobacter aceti*** is the bacterium discovered by Louis Pasteur, which turns wine (ethanol) into vinegar (acetic acid). It is part of the *Acetobacter* genus that all produce acetic acid. Unlike the LAB they depend on oxygen and live off a variety of sugars they can ferment, such as flowers, honey, fruits and in soil. Also unlike the LAB they are quite mobile, can quickly form layers and have adapted a resistance to acid so they can continue to live in the vinegar they produce. When the microbe *Acetobacter* annoyingly floats into an open bottle of red wine that you forgot about, it ingests the alcohol and, using two of its enzymes, converts it to acetaldehyde and then to acetic acid as a by-product. It then excretes the acetic acid into the wine and this slowly converts it to vinegar. *Acetobacter* needs oxygen to work, which is why you can store an open bottle of red wine for a few days if you use a vacuum pump to deprive the microbes of oxygen, or if you store it in the fridge, which buys you more time as the drop in temperature slows down any microbe activity. As well as vinegar production it is used in many industrial processes including biofuels.

***Bacillus*** is a rod-shaped bacterium that is a bit of an all-rounder – it can live with and without oxygen and, as an omnivore, it can eat carbohydrates, proteins and sometimes fats. It is tough too; when in trouble, it forms spores by folding over itself, cleverly doubling its cell wall. This makes it resistant to heat and cold, allowing it to persist for decades or centuries. Its main use is transforming the taste and texture of food. It can break down the tough proteins of soybeans to produce the soft, slimy and smelly natto with its distinctive flavours and provides extra vitamins like vitamin K2.

***Bifidobacteria*** are branch-shaped static bugs that don't like oxygen and eat carbs. They are the commonest microbe in the infant gut, enjoying body temperature and digesting breast milk oligosaccharides that help the baby to thrive. They are versatile and can produce both lactic acid and acetic acid in small amounts. They are also good team players, cooperating with many different species, such as the *Lactobacilli* and *Streptococci* in a range of fermented dairy foods to enhance flavour as well as health benefits.

***Propionibacterium*** is an example of a real specialist fermenter. It is key to Swiss (and alpine) cheese-making and makes the holes in Emmental cheese. It relies on other microbes surrounding it to produce lactic acid, which it feeds off; in turn it produces propionic acid and acetic acid to aid fermentation. When well fed, it also eats fats and produces carbon dioxide, which gives Emmental its distinctive holes, and fatty acids, which are responsible for its nutty taste.

***Enterobacteria*** Everyone needs to know a bit about bacterial food poisoning, which happens when food is contaminated with microbes that cause us to have symptoms of nausea, vomiting and diarrhoea – our bodies' way of getting rid of these so-called pathogenic microbes as soon as possible. The common causes are a group of microbes called the *Enterobacteria*, a broad group of bacteria named because they mostly live inside us (entero means inside) as well as in many foods. *Salmonella* is a well-known example, commonly found on meats like raw chicken (25 per cent of chicken meat in the US and 5 per cent in the UK), on eggshells (unless cleaned with bleach or chlorine) and on 2 per cent of raw minced beef. It can appear, rarely, in dairy and contaminated processed foods, as well as on some raw fruits and vegetables – often through contact with water, manure, soil, compost or humans.

*Enterobacteria* are active and mobile, often with little tails; they like low levels of oxygen, dark conditions and – despite a few unpleasant ones like *Salmonella* – they are often helpful; many can produce lactic acid to speed up the very early stages of fermentation. They can also convert nitrates to nitrites in our guts and some eat dietary fats. Some species, when feeding off fermented milk or vegetables, have a nasty habit of producing off smells or spoilage. As well as *Salmonella*, other well-known members of this group include *E. coli*, *Klebsiella* and *Shigella*. You don't want too many of these guys as they can give you diarrhoea and vomiting. The good news is that they can't cope with too much acidity, so just like heating them with cooking, if the fermentation proceeds normally, they will have died off before you eat the fermented food.

## Yeasts

Yeasts are around five times larger in size than bacteria. They are part of the broad fungi family that includes yeasts, moulds and mushrooms and are neither animals nor plants. They are key to the successful fermentation of many foods we have been eating for millennia, such as bread, wine and beer, nearly always in combination with bacteria, with just the proportions adjusted. They are a tough, flexible breed and can happily live with or without oxygen and are resistant to hot and cold temperatures as well as acidity, although they prefer working between 15 and 30°C. They are seasoned drinkers and can survive at alcohol levels up to 16 per cent. They can produce alcohol from sugar, with the by-product of fizzy carbon dioxide, which is also what makes bread rise. They produce many flavour compounds and a dried version called nutritional yeast is a good vegan alternative to Parmesan cheese as they both contain glutamates.

They are single-celled organisms that join together to live in large colonies with over a thousand known species. *Saccharomyces* is the most famous fermenting species and is present in nearly all fermented and alcoholic drinks as well as most kefirs; they are also found in our guts. In the presence of oxygen they grow well and reproduce, but they don't produce alcohol until they are deprived of oxygen in the fermentation phase. They can't survive long without oxygen, however, and like to be stirred to produce aeration, which keeps them going for many generations. Although our ancestors wouldn't have been able to see them before microscopes were invented, the word yeast in English, French and German originally meant zesty, bubbling, raising or lifting, so they certainly knew what they were capable of.

Many of these live inside our intestines happily and play a role in keeping us healthy. The three main types of yeast found in the vast majority of normal human intestines are *Malassezia*, *Candida* and *Saccharomyces cerevisiae,* otherwise known as brewer's yeast, which over 60 per cent of adults have inside them. There is a very rare condition called 'auto-brewery syndrome' caused by an overgrowth of natural brewer's yeast in the small intestine, which produces alcohol after eating food – an impressive excuse for failing a breathalyser test. Apart from occurrences of yeast overgrowth when we are immunosuppressed or take antibiotics, often causing vaginal thrush in women, yeast infections of the gut are rare. We all have small amounts of *Candida* in our mouths and intestines, which is perfectly normal and healthy as they help maintain our immune protection. Despite knowing this for at least twenty years, there is a huge and largely bogus health industry focused on convincing us that our guts are overrun with *Candida* and other yeasts and that we need expensive treatments to eliminate them.

Most yeasts are floating around in our natural environment looking for something to eat. Yeasts can help ferment most common foods and drinks, including bread, kefir, kimchi, krauts, kombucha and soybeans. Those involved in fermenting beer or bread convert sugars into alcohol and often carbon dioxide, but only if there is plenty of oxygen present. They also helpfully produce an enzyme called invertase that chops up sugars into smaller components like glucose and fructose, which it and other microbes such as lactic acid bacteria (LAB) can use. As the alcohol content increases, other microbes struggle to survive; above a certain level even the yeast itself dies off in a strange form of alcoholic suicide. Yeasts rarely work alone outside of industrial fermenting. In kombucha and most kefirs, lactic acid bacteria and *Acetobacter* often coexist with yeasts and work best together as a team to produce the chemicals to fight off other species. The networks of microbes in kombucha and in kefir produce a tough sugar structure, or biofilm, that holds the whole community together, making them a very tough opposition against changes in temperature or invading microbes. While many natural yeasts enjoy making alcohol, a few bacteria can also perform the same trick from sugar (such as *Zymomonas*) and there may be other alcohol lovers we haven't yet discovered.

***Aspergillus oryzae* or Koji mould** is a form of yeast and deserves special mention as the most important microorganism in East Asia. In Japan it has cult-like status. It is key to the fermentation of soybeans to make soy sauce, miso, sake and shochu rice wines and natto. It is usually first nourished on boiled white rice, where it grows and converts the starch into sugar, before being added to the soybeans or other plants to ferment. It stands out as being a powerful and versatile fungus that has the ability to dramatically transform the tastes and flavours of food and drink, releasing new aroma chemicals. It can cope with anaerobic conditions but generally prefers some air, ideally at 25–30°C, and moist and humid conditions, preferably in the dark. This mould often works with other microbes such as *Lactobacillus* when producing miso paste from soybeans. Other famous food moulds include *Penicillium roqueforti*, which is introduced via needles into Roquefort cheese to provide oxygen for the mould and give it its characteristic blue streaks. Genetically, fungi are more similar to humans than they are to plants; many edible fungi like mushrooms are either very good for our health or, if we pick the wrong ones, can kill us. Koji mould can also be added to nearly any plant to slowly ferment it: I have some roasted sweet potatoes slowly 'cooking' for eighteen months in koji, as well as soybeans that have been 'cooking' in salt and koji for six months. It transforms the taste and flavours from the mundane to the extraordinary.

## Viruses

Don't be alarmed. Viruses are everywhere in our food and, although occasionally known to cause medical problems, they can often be beneficial. We have heard of Norovirus food outbreaks as well as those in raw seafood, but this is rare in fermented foods. There are ten times as many viruses as bacteria in our guts and in most foods. These small phage viruses (phage means eating) evolved to be predatory parasites and feed off bacteria as well as larger yeasts like *Saccharomyces*. Imagine that every time you eat 10 billion bacteria (CFU) in your yogurt, you are also eating 100 billion phages.

We used to assume they were only harmful, with infections killing off starter cultures in dairies, vineyards and fermented soy factories and generally making life difficult. The commonest family of food viruses are *Caudovirales*, which in science speak means virus with a tail. They use their tails to attach to their favourite prey, a lactic acid bacteria, and then make it explode. In rare cases, they can cause a starter culture to fail if they overwhelm the starter microbes, but more often than not, they are beneficial, helping the fermenting process by maintaining the balance of species and ensuring that no species over-dominates.

Once inside our guts, viruses can also act to our benefit by killing off pathogens and helping send signals to the immune system in the gut wall. Sometimes these viruses help change the taste and smell of the food. One study of cocoa bean fermentation found that the phage virus that destroyed the LAB helped liberate enzymes that were key for increasing aroma chemicals.

Viruses are everywhere, and there is little you can do to influence them, but if your starter for fermentation is diverse in terms of species, it is unlikely they will disrupt the process and most likely will improve it.

## SCOBYs and mixed communities

A SCOBY is not a single microorganism; it stands for symbiotic culture of bacteria and yeast. It is a rubbery, slimy, disk-like culture that can trigger fermentation composed of multiple bacteria (at least twenty to thirty), including lactic acid and acetic acid, producing different varieties and several yeast species living together in a colony. A SCOBY is mainly used to make kombucha, but a related colony can rapidly make vinegars, also called a vinegar mother. The thirty or so microbes work as a team in kombucha, to eat up the sugars and use the tea leaves to create extra flavour molecules as the acidity increases and they produce natural fizz. To protect themselves they form a slimy coating called a biofilm that eventually allows itself to reproduce. They like mild temperatures (18–30°C), they don't like direct sunlight and prefer an acidic environment. Although most of the team in the blob are anaerobic, it does like an occasional draught of air to take away the excess gases. They can live for years in your fridge and will continue to grow thicker and can become so tough they can be made into leather-like clothes. Other examples of mixed microbial colonies are kefir grains, which the yeast and bacteria species produce and then use as a solid structure to live on and offer them protection (we will talk more about this later).

## Biotics – the living and the dead

There are many different terms used by food producers and scientists to describe foods and microbes. To add extra confusion, many of those definitions changed in 2021 following an international agreement. The term 'biota' means life in Greek, and 'biotic' has been used broadly to describe foods or additions to foods that can improve life (our health) but usually implies living organisms. In Europe and to a lesser extent North America, it is surprisingly hard for manufacturers to use scientific terms like probiotic, without enduring years of paperwork and expensive clinical trials. Most manufacturers get around these outdated rules by simply adding a tiny amount of magnesium, calcium, copper or zinc to their product, to make spurious health claims. These claims are based on thirty-year-old rules, and most experts now agree they are worthless, based on original evidence that was flawed. All that these regulations do is help the big food companies make simple short-cuts and ignore the needs of the consumer.

**Probiotics** are defined as live microorganisms that can benefit your health when you consume them in large enough quantities. They appear both in fermented foods and as capsules containing the microbes, as individual strains or as mixes. In general, fermented foods provide a greater range of microbes than found in most probiotic mixes, though science in both areas is changing fast.

**Prebiotic** is a general term for compounds that feed your gut-healthy bacteria (like fertiliser for plants) and are shown to have health benefits. For instance, there are types of dietary fibre abundant in plant-based foods that will nourish your gut bugs. Within fermented foods the carbohydrates that accompany the probiotic microbes in the food (supplying the microbes with nutrition) could potentially be considered prebiotics. Examples might be the cabbage in sauerkraut, or the lactose in milk yogurt. I found out first-hand how hard it is to label a product as a prebiotic even when it clearly is. With ZOE in 2024, we launched a prebiotic mix of thirty-two freeze-dried high-fibre plants and mushrooms (ZOE Daily30+) that can be added to regular meals. Before launching we performed a large, randomised trial on 399 people compared to a placebo and it significantly improved the gut microbiome and many symptoms in six weeks. Although in scientific papers and outside Europe we could refer to it as a prebiotic, in Europe we were not allowed to use the P word on the packaging or in marketing and bizarrely couldn't talk about the trial on the label or in marketing. Instead we had to call it a whole food supplement.

**Synbiotics** is just a word for a mix of prebiotics and probiotics where the addition of the prebiotic is shown to benefit the probiotic (e.g. they act synergistically), the idea being the prebiotic keeps the bacteria happy and nourished for longer. So, if you dipped some high-fibre fruit, like berries, into natural yogurt, you could consider that a homemade synbiotic. Manufactured synbiotics often contain some form of manufactured fibre like inulin with a probiotic in a capsule or powder, making them more expensive. Sauerkraut or kimchi are great examples of natural fibre synbiotics that existed long before these definitions.

**Postbiotics** are a new but important concept. They can be any mixture of whole or fragments of dead microorganisms, or the compounds or even the liquid they produced when alive. As examples, postbiotics could include parts of bacterial cell walls or metabolites that bacteria produce as they break down the fibre in your cabbage. Importantly, to be a true postbiotic, it has to benefit health in some way. In 2021, the International Scientific Association for Probiotics and Prebiotics after much heated discussion defined a postbiotic as a 'preparation of inanimate microorganisms (and/or their components) that confers a health benefit on the host'. They rejected other suggestions including parabiotics, or my favourite 'ghostbiotics'. You can now purchase a few commercial postbiotics in supplement form that have been clearly shown to work and they will become more common in the future. So it turns out that even dead microbes (bacteria and yeast) found in fermented foods are beneficial to our health.

This isn't so strange when you think about how many vaccines work. Vaccines can be both alive and dead, or at least inactivated. The dead vaccines containing dead parts of a virus (such as polio virus) or a bacteria (such as those that cause meningitis) still stimulate a protective immune response of antibodies that is useful and often produces fewer side effects such as fever or tiredness. They are also cheaper to use and have a long and stable shelf life and can't infect others like live vaccines. The disadvantages are that the immune response is generally not as powerful so protection is shorter and top-ups are needed. The fact that probiotics and microbes in fermented foods seem to have similar actions is good news. Even mistreated microbes that have been neglected, overheated, pasteurised, frozen, starved or overfed on sugar can still provide some benefit. Importantly the studies show that the mechanisms and effects are different to the live bugs, which still confer the greatest health benefits. But this whole new research area has meant we need to reorganise and clarify our terminology of fermented foods and microbes in this space, which could have big impacts on food labels, health claims and consumers.

Postbiotics and healthy dead ferments are real and are here to stay. Time will tell what percentage of dead microbes found in food have a measurable benefit, but enough evidence has been found already to make this an exciting field of research. Using freeze-drying and heating we can create a whole range of medicinal foods that could have major impacts on health, especially given the advantages: their relative safety to use in sick, young, old or pregnant humans, their long shelf life, and their ability to survive the human gut.

*

Now that we've looked at the basics of what fermentation is and how it works, next we'll explore why it is so good for our health and how we humans have actually evolved to like the taste of fermented foods.

### *What is fermentation? in five*

1. Fermentation is the transformation of food by microbes, changing its structure and taste. This can occur outside and inside our bodies.
2. There are thousands of species of bacteria (and a few fungi) that naturally live off food, making up 3 per cent of our total gut microbes in adults and over 50 per cent in newborns.
3. When you add salt or sugar or change oxygen levels, the specialist fermenting microbes dominate and produce acids and gases, which kill off competition.
4. Many foods are produced by fermentation, but then cooked or pasteurised, which kills the microbes (sourdough, beer, coffee, commercial sauerkraut).
5. Fermented foods are a mix of probiotics (microbes), prebiotics (food for microbes) and postbiotics (dead microbes) that all confer specific health benefits.

PART TWO

# Why Do We Ferment?

## Fermenting around the world

We have probably been eating fermented foods for most of human history and evolved to enjoy the estimated 5,000 different varieties out there. Fermented foods are central to the cuisines of vast swathes of the planet, whether it's yogurt in India, fermented vegetables in China, miso and other soybean products in Japan, kimchi in Korea and Vietnam, grain porridges in Africa, sauerkraut in Germany, or kombuchas and kefirs in Eastern Europe and Russia. Fermented fish (called garum) was key to ancient Roman cuisine, and eating fermented grain as porridge was common around the globe for thousands of years. Every continent has its own fermenting history, with variations depending on local conditions and how early animal husbandry was adopted. The Middle East, India and Europe developed dairy fermentation early on, whereas China, Japan and Korea focused on fermenting vegetables, rice, fish and soy, and Africa focused on fermenting its many varieties of grains and root crops. Fermentation not only changes the structure and chemicals of the food and drink, but importantly it changes its taste and health properties; for our ancestors, vitally, this made it safe to consume for longer.

While there is clear evidence of humans having the tools for fermenting as far back as 4000 to possibly 12000 BCE, we may have been using crude forms of fermentation since we split from apes over 10 million years ago. It has been proposed that before our discovery of fire, our discovery of fermented foods and their extra nutrients kick-started the rapid growth of human brains at the expense of our guts. A detailed study of different primate species (our ancestors) found that 15 per cent of those studied had clear evidence of seeking and eating fermented fruits, usually with some alcohol content. This close association with fermented foods in early man may have allowed the evolution of genes for processing alcohol, which has been shown to have mutated under selection pressure long before the spread of agriculture, maybe up to a million years ago. Our drive to seek out these fermented fruits could be explained by our relative lack of resident species of microbes compared to other primates, so we needed to consume ferments to counteract this. It seems likely that we evolved specifically to consume fermented foods as a regular part of our diet, something many of us seem to have forgotten.

The British – and by extension the settlers of the old British Empire, the US, Canada, Australia and New Zealand – have relatively few fermented food traditions that are still passed on, apart from cheese and beer. The reasons for this are unclear but food historians suggest that the early adoption of the Industrial Revolution after 1750 and the rapid drive to urban living may have had an impact, and recipes and traditions were lost that probably started even before the Romans brought their culinary skills and fish sauces to Britain. By the eighteenth century there are few records of fermenting in the parts of Britain where most industrialisation occurred; in more isolated areas, fermenting traditions survived for much longer.

Ireland, which remained in feudal and near medieval conditions longer than Britain, kept its local traditions, however. There are reports in 1690 of recipes for twenty different types of fermented milks with the favourite apparently being the sourest. One of these, probably the simplest, was called 'bainne clabair', which made it to the US via Ulster immigrants as Bonny Clabber and is still found in the Appalachian mountains, where 'clabbering' raw milk means leaving it out at room temperature to let it go sour without it curdling. In the highlands of Scotland, the tradition of eating fermented oat kernels called sowens as a porridge and baked as scones was very popular until about the 1960s (it is still sometimes eaten as a New Years' treat). In Industrialised England, by contrast, the milder winters and lack of mountainous areas also meant that the need to ferment for survival was less pressing than in other regions.

There are two competing reasons that our human ancestors became interested in fermentation: preservation and pleasure. Fermentation allowed them to keep drinks like mead, ale and wine for months without spoiling, at a time when water could cause diseases.

People with cattle could transform milk, which goes sour and mouldy after twenty-four hours without refrigeration, into yogurts, kefirs or cheese that could last weeks or months. We adapted to like the sour tastes that were associated with vitamins and nutrients and, of course, like some of our primate ancestors, we had a particular love of alcohol. This would have first appeared when fruit rotted at the bottom of a tree, or grain for making bread ended up making beer instead because of a chance encounter with a natural yeast.

## Aversion and risks

People in countries where fermented food traditions remain strong are used to the sour flavours and slimy textures that microbes produce. To cultures that no longer teach the benefits of fermenting, there can be something distasteful about the process. We have a natural aversion to moulds as the colours, hairy strands and textures send warnings to our brains that say – avoid!

Of course, many moulds, fungi and mushrooms are harmful to humans, but most that grow on our food are not. My mother, who grew up in wartime Australia with killer spiders and outside toilets, would always scrape off mould from cheese and serve it to my brother and me, who were squeamish, having been brought up in the more sterile sixties and seventies Britain. I don't remember ever having food poisoning from a home-cooked meal as a child or hearing about friends having that experience. So why are we so fearful of the fermenting process? In fact, what happens in a fermentation jar, whether it contains cabbage and salt or tea and sugar, is mirroring what happens every day in our large intestine as trillions of microbes attack and transform our food into healthy chemicals. To understand fermenting is to understand gut health, so it is crucial to optimising your long-term health and preventing disease.

While reports of infections due to fermented foods are very rare in high-income countries, if you are travelling in lower-income areas of Africa, be warned that unpasteurised camel and cow milk ferments can carry nasty strains of *E. coli*. There are also a few reports that fermented vegetables can sometimes get contaminated with unwanted microbes and cause food poisoning. Some large Korean outbreaks have been traced back to water contaminated with *E. coli* that was used to add to the salty brine. These kimchi scares are due to a few isolated outbreaks, which need to be balanced by the near universal consumption of this largely homemade product in Korea, Japan and China by billions of people. US public health experts have, in classic risk-aversion tactics, warned against home fermenting, but forgot to tell us that food poisoning from US restaurants carries a far greater risk, especially with *Salmonella*.

Another commonly reported potential problem with eating ferments is due to histamine allergy. Many people claim that cheese gives them nightmares or a range of other odd headache symptoms grouped together as a 'cheese reaction'. In 1996 this cheese reaction was linked to high levels of tyramine and histamine produced by acid-loving microbes present in most fermented foods. Although a few people are genuinely affected, for most it is a subjective, imagined reaction, such as the MSG headaches falsely claimed to be caused by Chinese food of the 1980s. Many people blame Stilton in particular for poor sleep, though as it's usually accompanied by red wine or port, this may be a convenient scapegoat. Levels of tyramine and histamine are increased the longer the cheese matures, increasing risk, although the balance of microbes used in the starter can now be manipulated to produce less. Rarely, other fermented products, particularly sauerkraut, but also chocolate, salamis, fermented pickles, wine and beer can cause similar problems in particularly sensitive people, for reasons we don't yet properly understand but are likely linked to their microbiome.

## Health benefits – the latest science

I always thought that dead bacteria were useless, and that commercial pasteurised or heated microbial products with yeast or bacteria were essentially a con, along with probiotic supplements kept in poor conditions for too long. It turns out I was wrong. New science has once again shown we were underestimating the complexity of microbes, but it takes a long time to change a scientific dogma. In the past, heat-inactivated microbes were considered inert and useless, so they were often used as placebo controls for studies of the active products. But scientists have for a while noticed the 'inert' placebo control arm often appearing to be doing something strange to the trial participants. The first few studies that produced these unexpected results were dismissed as chance findings or errors. In 2010, a Japanese commercial study of eighty elderly people randomised them to placebo or heat-inactivated *Lactobacillus pentosus* strain b240 for twelve weeks and found it increased levels of the natural antibody IgA in saliva, which could potentially help fight infections. They followed this up with a bigger randomised study of 280 adults and found a 40 per cent reduction in common cold infections, but being a commercial study performed in Japan it had little impact. In 2011, some food chemists used a *Lactobacillus* specialised for kefir (*L. kefiranofaciens*) in a trial and found that the pasteurised microbe, used as a placebo in the control arm, still had immune and anti-asthma effects. The finding was only in mice, and in a minor food chemistry journal, so again little attention was paid to it. But gradually science and opinions have begun to change as the evidence became overwhelming.

A key review, published in 2020, summarised seven randomised controlled trials from more than 1,700 children under five for treatment of diarrhoea. Heat-killed *L. acidophilus* (a LAB often found in fermented dairy) significantly reduced the duration of diarrhoea compared to placebo. Heat-inactivated *L. paracasei* CBA L74 also reduced the risk of developing sore throats and when added to milk formula for infants appeared to improve gut health and produced an immune response in the gut (a rise in IgA antibodies). A large German study of 443 patients with irritable bowel syndrome found in 2020 that a dead form of *Bifidobacterium bifidum* (another common species in fermented foods) reduced symptoms after eight weeks. Other small studies with dead traditional probiotics, such as *L. lactis* and *Bifidobacterium*, have confirmed these were not freak results.

Heat-killed probiotic microbes are one thing, but what about those in food? I have always wondered whether all the microbes in sourdough bread had any effect after baking. So I was delighted to read a 2023 study that took the obscure microbe *Companilactobacillus crustorum* found in sourdough (and koumiss fermented milk) and, after heat-inactivating it, fed it to mice with induced colitis (gut inflammation). The study found that it improved their gut walls and reduced inflammation. Most of the probiotic microbes that have been found to have health effects in animals and humans when dead are also found in similar forms in fermented foods, which are a complex mix of microbes. As well as dead kefir studies in mice, we now have some recent pilot studies in humans, where consumption of the dead milk kefir with selected artisan microbes resulted in a reduction in lipid levels and inflammation that did not occur when a commercial milk kefir control was used. We are now seeing that heat-killed kombucha has some antimicrobial activity and small sauerkraut studies in humans found that the pasteurised arm had the same benefits as the live arm for irritable bowel syndrome (IBS) sufferers.

The science is taking us even further into new territories, including into the next generation of human gut probiotics not yet found in food or ferments such as *Akkermansia*, which first appeared as a possible protective microbe against obesity in our twin studies in 2014. This fascinating bug has a consistently higher presence in people with more lean muscle mass (metabolically healthier) and increases after long overnight fasts, when in the absence of food it nibbles on the sugars (glycans) making up the protective mucous layer of our thin gut barrier, protecting our blood from microbes. This keeps the mucous layer neat – like a freshly mown lawn in our gut garden – and so less prone to leakage and inflammation. *Akkermansia* has some blood glucose lowering properties and possibly some mild weight loss effects. My Belgian colleague Patrice Cani was surprised to find in 2019 that when pasteurised, its anti-obesity and anti-diabetic effects on mice, and then humans, were actually greater than when it was alive.

We might now expect a weaker or similar effect of dead compared to live microbes, but how to explain the greater benefits of dead *Akkermansia* microbes compared to live ones? Some clues come from its array of proteins on the cell wall, but it also secretes a novel protein (P9) that stimulates the anti-hunger gut hormone GLP-1, the same target of the anti-hunger obesity drugs Ozempic and Wegovy. Another protein on its surface has been found to help with the integrity of the gut barrier. Finally we now know that microbes like *Akkermansia* alter energy and blood sugar control via a system (called PPARgamma) and it doesn't matter much if they are dead or alive. Depending on where you live, you can now buy both the live (US) and dead (Europe) product commercially – Danone have recently bought rights to the dead version – and the effects may be slightly different.

Unlike taking a selection of shop-bought probiotics, fermented foods are not just a collection of random microbes: they are a tight team working together to use all the chemicals they produce without waste and maximise their collective survival. For example, the microbes in fermented milks and teas produce exopolysaccharides, which are firm sugar-based structures (grains and SCOBYs) that protect them and optimise the microbes' survival and biological actions. A growing number of lab studies show that these inert structures, rather than the live microbes themselves, enhance features of the immune response, such as reacting to vaccines, or by reducing inflammation in the blood and improving the gut barrier in human trials of a killed *Bifidobacterium* probiotic versus a placebo in middle-aged women. Other studies have shown that so-called fermentates (the dry powdered form of fermented foods) have a key role interacting with viruses and immune dendritic cells. Others have found these structures could have a metabolic effect, i.e. by lowering blood pressure or reducing obesity and cholesterol in rats, in some ways similar to statins. This means that some fermented foods could help someone with a poor response to statins, if the successful mouse studies can be replicated in humans. Some studies are investigating whether these structures could even be useful as anti-cancer agents.

Until recently there has been a lack of detailed or large-scale studies of the health benefits of fermented foods. Together with the ZOE team, which did such great work launching citizen science research around the impact of Covid-19, we organised a project to explore how people not familiar with ferments would adapt to eating three portions a day of different types. People were quick to volunteer, and we rapidly recruited nearly 10,000 volunteers with a special app where they could report their daily symptoms in real time. We selected people aged eighteen to ninety who had no special allergies and were currently eating less than one portion of ferments daily and asked them to increase this to three portions of different types (including yogurt, kefir, kimchi, kraut, kombucha and water kefir (tibicos)) for the next two weeks. One serving was 175ml of a drink and a quarter of a cup for krauts, etc. At baseline they were having an average of half a portion a week; after fourteen days, this had risen to three portions a week on average – a sixfold increase. We also asked them to score daily (on a scale of 0–10) their mood, energy, hunger and bloating scores.

A total of 5,472 people (mean age 64 and BMI 24) managed to log foods every day and also increase their ferment intakes three-fold. Another 1,021 (mean age 63) had logged their foods and were taking extra ferments but not as consistently and did not increase intakes as much. By the end of the short study, around 50 per cent of people had seen benefits to mood, energy, hunger and bloating and only a minority had felt worse or had experienced increased bloating. The group that tripled their intakes consistently reported more improvement across all symptoms compared to those that increased less or inconsistently, suggesting a dose-response effect. The study did not have a dummy or placebo arm, which is virtually impossible with food, and some of the results could have been due to well-known placebo effects. But the fact that the less compliant group of 1,021 men and women did not report as much benefit, supports a real *biological* rather than simply a *psychological* effect. This unique data drawn from people's real lives, rather than small groups of highly selected young volunteers (usually males) in labs, gives us an exceptional insight into the huge potential benefits of ferments for the general population. Many volunteers wrote to us with their personal anecdotes of how the experiment had totally changed their attitude to these foods and gut health.

The benefits of fermented foods are so wide-ranging, that it might be helpful to break them down into different categories.

## Digestion and gut health

The process of fermentation not only breaks down tough fibrous structures into potentially edible foods, but it also transforms the chemicals in the plant in terms of taste, reducing bitterness or sourness and increasing the complexity of aromas and volatile chemicals. The act of fermentation usually increases the digestibility of the food, breaking it into smaller pieces, which our bodies can deal with more easily without triggering an immune response, and extract more nutrients, producing a greater variety of healthy chemicals. This means that like cooking, fermented foods are broadly easier to digest than their unfermented forms (examples include sourdough bread for gluten intolerance or yogurt for lactose intolerance). Some studies report initial bloating when starting to eat fermented foods for the first time, but most show a reduction in gut symptoms over time. Our own ZOE Ferment study showed that on average, after two weeks, 42 per cent of people had fewer gut health issues, such as bloating and constipation, but 24 per cent reported more bloating symptoms. About one in five people suffer from some form of irritable bowel syndrome, which presents with constipation or diarrhoea and bloating and pain after meals. A lady called Sharon wrote to me: 'For as long as I can remember I have experienced periods of abdominal discomfort and bloating.' She used to only eat yogurt but realised she needed to embrace other ferments much more; this helped to eradicate her bloating and has certainly improved her constipation. Alison said, 'After suffering from IBS in various forms most of my life and trying just about everything there is I tried kimchi. Massive improvement! Ninety per cent better, all symptoms gone, but they return if I stop eating it.' A forty-five-year-old lady called Claire wrote to me, recently diagnosed with diverticular disease. Instead of taking laxatives, she started three ferments per day and extra water; within a few weeks, she had no pain or bloating and no constipation. Robert is another convert to ferments who now makes his own kefir. He wrote to tell me that he used to suffer badly from oesophageal reflux (heartburn) and that he was able to wean himself off his drugs (omeprazole). We have to remember that responses are personalised, and some people may need to increase their ferments more slowly to reduce bloating.

A major impact of fermented foods is clearly on the gut. Gut health is difficult to define and depends on your perspective. From a **metabolic** viewpoint, a healthy gut means ensuring that the intake and absorption of food and corresponding hormonal hunger signals are well balanced for the body's needs, preventing obesity and diabetes. From an **immunity** viewpoint, the gut lining is another crucial player in a healthy gut. A thin layer of cells and a thicker wall of sticky mucus separates the billions of microbes in the gut from the blood vessels that supply our gut and enter our circulation. Fermented foods and fibre play an important role in maintaining this barrier. When we lose the mucus layer, the tight structure breaks down and the system leaks, causing inflammation. This may be a reason fermented foods are helpful to many people suffering with colitis or IBS, with many people writing to tell me they were able to reduce or stop their medication. And finally, from the viewpoint of our **nervous system**, a healthy gut means normal intestinal muscular contractions and regular soft stools and trips to the toilet, avoiding bloating and cramps and IBS. Another (simpler) way to define gut health is a healthy gut microbiome . . .

## Balancing the gut microbiome

Ferments improve microbiome health by stopping overgrowth of pathogens and encouraging greater species diversity. Microbial diversity or richness is now used scientifically as the best overall measure to explain the wide differences between people's health. Microbial diversity has been shown in countless studies to be a crucial parameter of health, with low diversity associated with every common western health problem, from depression to diabetes, allergies to autoimmune disease, and poor response to anti-cancer drugs. As well as increasing diversity and the richness of the community, the probiotic components of fermented foods will also increase the amount of healthy bugs and reduce the amount of unhealthy bugs. In several recent ZOE studies we have shown that this good/bad ratio of 100 common microbes is the most important indicator for heart and metabolic health, and this ratio is proving to be a more accurate measure than diversity. We also showed (in the ZOE biome clinical study) that taking a probiotic containing *Lactobacillus* for six weeks improved this good/bad ratio compared to placebo, although not as much as our prebiotic mix of thirty freeze-dried plants. Fermented foods are usually a mix of probiotics and prebiotics so they are likely to be even more powerful in terms of gut benefits, as shown by a number of small-scale human studies.

## Hunger and weight maintenance

Ferments have been linked with weight loss in a number of studies, but the data has generally been poor quality and small scale. I regularly get told by people that just changing to regular ferments helped them lose weight. Most tell me that hunger is less and their weight loss is a slow, gradual process of maybe losing 5kg over a year, but this is highly variable between people. We now know that microbes play a role in hunger and satiety and some microbes can secrete hunger-suppressing chemicals like GLP-1 (the basis of drugs like Ozempic).

Some probiotic companies promote their products as a cheap form of Ozempic or Wegovy, but the effect of these chemicals in real life is as yet unclear and may have minimal effects in humans. All fermenting microbes produce other chemicals that can potentially help with weight loss by reducing inflammation, helping fat metabolism, shrinking fat cells and helping glucose metabolism. Several observational studies have shown that regular yogurt eaters have lower BMI than non-eaters, but this is far from proof, as many other factors are at play. Trials are better evidence. Our own ZOE Ferment study did show that hunger reduced significantly during the study in 52 per cent of volunteers and increased in 25 per cent. The greater the body weight at baseline, the greater the reduction in feelings of hunger. So there is growing evidence that ferments could reduce hunger in most people.

## Supporting the immune system

Nearly 80 per cent of all our immune cells are lining our gut, and they are there for a reason: to talk and respond to the chemicals in our food and our gut microbes. A well-functioning immune system is crucial to our health. It prevents allergies, infections, autoimmune diseases, cancer and – importantly – regulates inflammation, which is the regular healing response of our body to everyday stresses like bad food choices, emotional stress, pollution, obesity and a number of diseases.

We now realise that persistent inflammation at low levels that are barely detectable in our blood is very different to the short-term inflammation we need to repair an injury. This persistent low-level inflammation is associated with nearly every common disease such as osteoarthritis, back and chronic pain syndromes and most immune and mental health disorders. It also predisposes us to heart disease and strokes. Just a small amount of inflammation in our blood vessels makes them more likely to form fatty plaques, stiffening the arteries and leading to strokes and heart attacks. Similar effects in brain vessels can lead to dementia.

Inflammation is now linked to increased risks of all types of cancer, and the likely mechanism is that if the immune system is constantly fire-fighting and dealing with low-level inflammation, it has less time and fewer resources to deal with early detection of rogue cells that, if ignored, turn into cancer cells. Several people have written to me to tell me that they switched to fermented foods after a cancer diagnosis or a scare, when doctors said they were at high risk. Although possible, there is no solid evidence yet that they prevent cancers, but it seems a good insurance policy along with other anti-cancer drugs.

Inflammation was also a key risk factor for severe Covid. During the pandemic, we surveyed half a million people with our free ZOE app and found that those taking probiotics or eating fermented foods regularly had less severe Covid symptoms (by 14 per cent) in the following year than those taking vitamin C, zinc, garlic or nothing. This is not proof by itself because it was an observational study and could be due to other health habits, but combined with other data and the negative effects of other supplements, it is compelling. It is worth bearing in mind, given that it is likely that we will see another major pandemic in the next thirty years.

At the other extreme end of the scale, and in a landmark study, my colleagues at Stanford University precisely measured the immune response in eighteen people having five daily portions of fermented foods compared to another eighteen following a high-fibre diet. They saw major shifts in the immune cells and a significant decrease in nineteen inflammatory proteins (such as interleukin-6) in just a few weeks. This was the first study to clearly show the immune effect of ferments in humans but also showed that fibre benefits immune cells in different ways. This is important as many ferments are synbiotics, meaning they contain prebiotic fibre as well as probiotic microbes (krauts, kimchi, etc).

Lab-produced ferments made with milk powder have been shown to have very specific effects on the immune system, suggesting these could be useful future anti-viral therapies. Men and women have very different immune systems, but our results from the ZOE ferment study showed similar benefits in men and women. We also saw no major differences between the types of ferments consumed, although as people tend to mix them, it is very hard to separate their effects in studies.

## Managing inflammatory conditions

As well as having a key role in regulation and prevention of diseases, ferments can help people with existing inflammatory conditions such as inflammatory bowel disease, or autoimmune diseases like rheumatoid arthritis. Unfortunately, the studies so far have not gone beyond lab experiments on culture dishes or in small rodent experiments; proper clinical studies are nearly non-existent as they are expensive to perform, especially on hospital patients, and no one appears willing to fund them. Although small, studies of commercial probiotics in these same diseases have, overall, been positive, suggesting that fermented foods, which contain a wider variety of microbes, should work in theory too. Our ZOE Ferment study showed that 52 per cent reported an increase in energy. Tiredness is a common sign of low-level inflammation and activation of the immune system, so suggests a rapid action of ferments on the immune system. Many people suffering with conditions such as rheumatoid arthritis or ulcerative colitis have told me in person at my lectures how ferments have improved their energy levels and reduced the fatigue that is so common in these conditions.

## Managing cancer

There is increasing interest in using diet and ferments in the treatment of cancer, which is now recognised as mainly an immune condition, caused by a failure of the immune system to kill the malignant cells early on. Cancer cells are inflammatory and change their local environment to survive. Sometimes you find unusual microbes on the surfaces of tumours that can act as early markers. Many studies and meta-analyses of epidemiology studies have linked yogurt intake to a 20 per cent lower cancer risk – especially of the colon. But this was associated with finding *Bifidobacteria* on the tumour surface when biopsied. So yogurt and other ferments may help in the early stages when the tumour has started to alter the local environment and weaken the gut lining. New immunotherapy drugs are transforming cancer treatments; I co-led large studies on advanced melanoma, which showed the vital importance of good gut health in having a life-saving response to these drugs. I generally recommend anybody starting cancer treatments with immunotherapy to maximise their gut health through their diet, including eating more fermented foods. Anecdotally patients have reported to me having fewer side effects on both chemotherapy and immunotherapy if they also take fermented foods, but this needs more studies.

## Mental health

Fermented foods have been shown to improve general mood, as well as specific conditions including anxiety and depression and possibly even ADHD. Many people report improvements in their energy and mental state anecdotally. Charlotte wrote: 'I was very stressed at work and was going through internal worries about wanting to look for a new job, etc. I incorporated kefir yogurt into my breakfast and quickly my gut problems were decreasing. I'm now in a position where I am far less stressed and I haven't experienced pain since Christmas.' Brain imaging studies have shown significant changes in the emotion centres after drinking fermented milks. Some studies suggest effects at least as large as antidepressants. Jacky is a GP in the UK and reported to me, 'I have incorporated a lot more fermented foods. I now make my own kefir and have this as well as kimchi or sauerkraut most days too. Since then I have reduced antidepressants from 30mg per day to 10mg three times a week and stopping completely soon.' A small clinical trial in Ireland found a combination of fermented foods and prebiotic fibres reduced perceived stress in twenty-four adults who improved after four weeks by 31 per cent, compared to twenty-one controls who improved by 17 per cent.

Our much larger ZOE Ferment study confirmed this with a rapid improvement in mood after increasing ferments in 56 per cent of people; in most people they noticed changes within a week. There was a clear dose-response effect, with those only slightly increasing intakes having only modest benefits. Until recently, we had only a rough idea why mood changes via the gut, as the brain and gut are not directly connected and the mind and body were always seen as separate. A prevailing view was that it was a failure of the gut to remove toxins, which caused these to leak into the blood and cross the tough-to-penetrate blood–brain barrier. Now we know of many new mechanisms. The first is that microbes in the gut can produce brain neurochemicals themselves (serotonin, tryptophan, GABA and dopamine as examples) that can travel to the brain directly. Other signals from food can pass via the vagus nerve or via signalling from the immune cells lining the gut.

For me the most exciting new mechanism is that gut immune cells can sense inflammation in the gut lining and send signals directly to the brain that tell it something is not right. The brain then responds by going into depression or anxiety mode. We now know that there is no real barrier to the gut and the brain communicating directly, whether via chemicals, immune cells or nerve signals. It makes more sense to think of the brain as a key organ that the gut communicates with. This has huge implications for how we view and treat mental disorders that might be more problems of abnormal reaction or sensing of the external environment than something purely internal to the brain.

In evolution, one of the earliest multi-celled creatures was the hydra. It initially evolved as a simple tube with food going in one end and waste out the other; eventually nerves developed around it to help the tube move. This suggests that the large complex nerve network around our guts, which we call our second brain, was actually our first, and we should pay it more respect.

## The microbial pharmacy

So how do fermented foods achieve all these diverse health benefits? What we have learned so far comes from piecing together many different strands of science, from studies of germ-free mice to clinical trials of probiotics in humans, to simple studies with volunteers eating more fermented foods. The key to this new science is understanding that microbes – whether in our food or our gut – are essentially mini pharmacies or chemical factories. We think it is the chemicals that microbes produce or the proteins in their cell walls that have the main effects, not their magic defensive powers like having a strong wiggly tail. But although we know that drinking a small shot of kefir will give you a few billion microbes in your mouth, only 10 or, if they're lucky, maybe 50 million will survive the journey to your large intestine. Once there, they will be outnumbered by the 100 trillion local residents by a factor of anything between 2,000 and 10,000 to one. You might imagine them as colonial invaders arriving in a foreign land and attempting to suppress multitudes of the indigenous residents. But in fact, there is no evidence that fermented food microbes act in this way. They are more likely to use stealth.

A theory I favour is that our fermented food lands upstream in the quiet waters of the small intestine, where the fermenting microbes can get ashore and reproduce with less risk of being wiped out. The odds are much better here, in fact a thousand times better, where they will be in equal numbers to the residents. This means the chemicals they produce can have a real impact. The wriggly 22-foot length of the small intestine is the perfect place to get lost and – as far as we've been able to uncover – it is full of food-sensing and immune cells.

In the last chapter we looked at the key fermenting microbes found in our food. Any non-fermented food that reaches the lower parts of the gut will still allow resident gut microbes to produce beneficial chemicals, but we think fermented foods boost this process. It is useful to look in detail at the main chemicals our fermenting microbes produce and how they might work.

Gut microbes produce thousands of different chemicals in their role as chemical factories. We still know very little about most of them. Some of the key chemicals are relevant to fermented foods, because either the fermenting microbes produce them themselves or act indirectly to stimulate production by the local residents. You don't need to remember all the names, but I want to introduce you to some of the key groups of chemicals, which show what a complex and remarkable process takes place inside our guts.

## Vitamins

The gut microbes make several vitamins, including vitamin C, essential for immunity and multiple processes; vitamin K, which is useful for blood clotting and bone health; and several B vitamins, such as folate and biotin, which are helpful for nerves. The fermenting microbes often produce vitamins as a by-product that helps other friendly bacteria nearby. *Streptococcus thermophilus* and other LAB found in yogurt and other ferments, and *Acetobacter* in kombucha and tempeh (fermented soy) can produce vitamin B12 and folate. Unfortunately, most of this B12 is produced in the large intestine and is not reabsorbed into the bloodstream. Sauerkraut is a great example, being a rich source of vitamin C and K, some vitamins coming from the plant itself and some produced by the fermenting yeast. The bacteria *Bacillus subtilis* found in natto is a source of vitamin K and kimchi, as a combination of many plants and microbes, provides vitamins A, B and C. Tempeh is a rare product that provides B12, even when cooked.

## Short chain fatty acids (SCFAs)

These nasty sounding chemicals are actually very healthy for our immune system and hunger. They are produced through the fermentation of dietary fibre and other complex carbohydrates (e.g. in cabbage) by bacteria. SCFAs such as acetate, propionate and butyrate play an important role in energy metabolism, providing microbes and the immune cells in the gut lining with an energy source. They also signal the body about hunger and fullness, possibly explaining why fermented foods tend to reduce hunger symptoms. They also play a vital role in maintaining immune function, reducing inflammation and gut barrier integrity. While they clearly work well within our bodies, their use as synthetic diet supplements has been disappointing.

## Neurotransmitters

The gut microbiome is involved in the production of several key brain chemicals, including serotonin, dopamine and gamma-aminobutyric acid (GABA). These molecules crucially influence our mood, behaviour and cognitive function. The ever-versatile LAB microbes can often produce GABA (which I think of as a mild form of natural Valium) when fermenting a range of different foods that contain enzymes to kick-start the process. It turns out that many traditional fermented foods such as cheese, yogurt, kimchi, koumiss, fermented beans and fermented fish sauce are a rich source for GABA-producing LAB. The LAB microbe *Weissella*, found naturally on many plants, has special strains that produce very high levels of GABA and is attracting commercial interest. Similarly, some other LAB microbes can convert tryptophan in food to compounds such as serotonin. This, when given as a *Lactobacillus* probiotic, has been shown to improve mood in rats and in a few human studies. Around 90 per cent of serotonin is made in our guts by microbes and has been associated with mood; it also influences how many antidepressant medications are believed to work.

## Secondary bile acids

Primary bile acids don't sound appetising but again are useful for us. They are released by the gallbladder into the small intestine to aid the digestion of fats. Gut bacteria play a key role in breaking these down and re-constructing them into another subtype called secondary bile acids. These acids then break down fats into smaller units for storage and energy. Secondary bile acids have a range of helpful physiological functions, including clearing lipids like cholesterol from the blood vessels and protecting our gut lining. Recently we and others performed studies that showed gut microbes also produce primary bile acids, previously thought to only be made by the liver. We found one bile acid (isoUDCA) that, when elevated, was a really good marker of inflammation and high lipids. Several common fermenting microbes can produce these helpful bile acids. Other studies in human trials have shown that the benefits of the Mediterranean diet on weight loss are strongly influenced by these bile acids. So eating ferments can improve your elimination of fats from the body *and* help protect your gut lining against leakage.

## Trimethylamine-N-oxide (TMAO)

This important compound could be the reason some people react badly to eating meat and are more likely to suffer heart attacks. It is produced through the breakdown of dietary choline and carnitine found in animal protein by gut bacteria. These gut bacteria convert it to an intermediary chemical called TMA and then other enzymes convert it to TMAO, which is the problem chemical. Elevated levels of TMAO have been consistently linked to increased heart disease and strokes, and overall mortality in many studies. But not everyone produces high levels of TMAO when they eat meat or fish. Because our resident gut microbes are very individual, our responses are very different. The microbes that we know produce harmful extra TMAO have mainly been from the Proteobacteria group, which includes *E. coli* and *Clostridia*, which are only rarely found in ferments, and so far other common fermenting microbes have not been implicated. In the future fermented foods could be used to reduce TMAO levels in meat eaters.

## Polyamines

You may have heard of spermidine as a novel anti-ageing supplement, but you may not know that microbes can also produce them. Polyamines, like spermidine, are small but vital molecules involved in various cellular processes, including DNA replication and cell proliferation that are key to preventing cancers and reducing ageing. Most animals and bacteria need polyamines to survive and there are high levels in mushrooms, peas and nuts. Some fermenting microbes, such as *L. reuteri* and *L. brevis*, also produce excess spermidine and putrescine (another polyamine). These chemicals then interact with gut lining cells to produce bioactive versions that get into our bloodstream. There are now some human studies showing that supplements have a beneficial effect and that levels decrease with age. Giving a combination of *Bifidobacteria* probiotics and the amino acid arginine has been shown to naturally increase microbial putrescine production, and in the future other combinations of fermented foods are showing promise, though we know little about the doses needed.

## Microbial juice

This is literally the liquid and all the chemicals that bacteria produce once you remove the cells (known scientifically as cell-free supernatants). Despite knowing little about them, these juices are already being sold commercially. Microbes are fed and grown in large tanks and will naturally produce a range of chemicals that leach into their surroundings. After removing the live bacteria, what's left are the juices. Some of this juice contains the cell walls of dead microbes, which have key proteins still intact. These proteins can still signal to the immune and sensor cells in the gut wall and other microbes to produce beneficial effects. This is why the dead cells and the microbial juices can be called postbiotics if found to help our health.

Lab studies showed that a microbial smoothie made with juices from five species of 'good' bacteria often found in fermented foods

(*L. acidophilus* and *L. casei* especially) had anti-inflammatory and anti-oxidant effects on the cells lining the intestines. Another lab study found juice from some *Lactobacillus* species could inhibit colon cancer or a disease-causing type of *E. coli* from invading gut cells. A few animal studies in mice and lambs have replicated these findings, but we still lack the hard evidence of human studies. Though early, these studies show us that the large numbers of chemicals produced by fermenting microbes have huge potential to help us. So you can now see why I use the analogy of the microbiome pharmacy and how ferments are some of these highly specialised medicines that can help us all.

## Personalising our microbes

One area that we've only touched on so far is the differences between our individual guts. The structure of the gut with its endless nerve and immune connections and complex gut lining has been cleverly shaped by evolution. While the structures and functions of the gut are quite stable over our lifespans and between generations, the length of the gut can vary between people by as much as 30 per cent. But it is in the community of gut microbes (our gut microbiome) that the greatest individual differences lie, as this dynamic community endlessly changes and adapts to different environments.

One of my rare 'Aha' moments in science was when I found out that identical twin sisters shared less than 25 per cent of their microbiome species with each other. It turned out to be even less when years later we did more fine-sequencing. In thirty years of studies of clinical markers of disease like blood and urine tests or blood pressure, I was used to finding strong similarities and had never found anything to match that level of difference. We are all born with sterile guts and they quickly get colonised after birth so we can digest breast milk and start to develop our own immune system. We know that our normal gut microbiome stabilises into our adult format around the age of four, with a core set of microbes that are the long-term residents, plus others that come and go like tourists. Some change with the seasons and some with our diets.

Studies of the microbiomes of Africans used to traditional local diets who were fed modern US diets have shown major changes in just two weeks, and Asian immigrants to the US markedly changed their gut microbe composition after eating American diets for a few months. As non-twins, we only share 10–20 per cent of our core microbes on average with fellow humans, and all of us have multiple rare strains that are unique to us. This is very different to our genes, where we share over 99 per cent with each other. What this means is that we all respond differently to foods, as our microbes will produce different sets of chemicals in response. In the ZOE PREDICT study in 2020, we showed that we have a tenfold different response to the same wheat muffin, suggesting that this extends to all foods.

No study has looked closely at personalised responses to fermented foods, but we have looked at prebiotic and probiotic supplements, and we can extrapolate from those results. The human trials show that although 'on average' there is a mean improvement in gut health, many people in the treatment groups fail to respond, while some respond very favourably. This implies that factors within an individual – likely the composition of their gut microbiome – determine the level of benefit fermented foods can afford.

We know that around 50 per cent of an infant gut microbiome is made up of food microbes, but this drops to only 3 per cent in adults. This suggests it must be difficult for food microbes to find a permanent home or a niche in the gut where they can have access to food and safety. A number of studies in mice, humans and other animals using probiotics and stool transplants from one mouse to another (called faecal microbiota transplants (FMT)) have shown that the ability of a newly ingested strain of microbes to take hold depends mainly on the existing strains that are occupying the same space. The more genetically similar newcomers are to the residents, the less likely they are to gain a foothold and colonise the gut, which means they will have less beneficial effect. This observation that our gut doesn't like accepting close family members is probably not personal, but likely because they have similar eating habits so they will compete for the same limited chemical food sources. Foreign microbes have different food tastes and so are less of a threat to the residents.

This so-called niche theory explains why some single strains of probiotics appear not to work in some people, why many faecal transplants fail and why single types of prebiotic fibre supplement are disappointing. The solution is to cover your bets and go for a full range of different microbes, hoping that at least some of them will differ from the residents. This is the basis behind new probiotics being sold as multi-strain, although this adds to the expense and some strains could be competing with each other.

Even better is to take a wide range of different fermented foods that are pretty much guaranteed to have probiotic microbes that differ from the resident strains in your gut and therefore more likely to find a niche. Even if the gut is crowded and they don't find a niche, if they are attached (as synbiotics) to some fibre, such as cabbage, they will last longer and produce helpful chemicals. If you consume three different ferments a day, you will probably be ingesting up to forty known microbe species, so you are definitely maximising your chances of good health.

### *Why do we ferment? in five*

1. Animals and humans have been eating fermented foods for millions of years, for pleasure, for health and for safe long-term storage.
2. Fermented foods help our digestion, microbiome balance, gut wall and immune system, control inflammatory conditions and improve our mood.
3. We evolved to eat fermented foods to help supplement our normal gut residents with microbes that have extra special properties.
4. Fermenting microbes have special properties, producing chemicals that interact with the rest of our mind, body and immune system, like vitamins, short chain fatty acids

(SCFA), neurotransmitters, secondary bile salts, polyamines and unidentified chemicals in microbial juices.

5. The gut microbiome is a community of trillions of bacteria, viruses, fungi and parasites that is unique to us all. Sharing the space with fermented food microbes makes the whole environment more healthy.

# PART THREE
# Ferments

Before we look in more detail at different kinds of fermented food and how to make them, let's try to clear up some of the confusion around ferments.

There are many fermented foods we recognise easily – yogurt, beer, wine, cheese and bread – but we often forget that some foods such as traditional salamis are fermented, as well as chocolate and coffee. Some of these products still contain live microbes when we eat them (most cheeses, yogurt, kefir and sauerkraut), while others have microbes that were key to their creation but are traditionally killed before we consume them, like sourdough bread, chocolate, beer or coffee. Others may be somewhere in 'no man's land' between dead and alive depending on how they are used or heated, like miso paste or soy sauce.

Many people get confused between pickling and fermentation, as products that traditionally used to be fermented with salt and microbes are often now pickled, meaning preserved with vinegar. A good example are pickled gherkins or cornichons, which are delicious but have few health benefits and lack the subtleties of taste you get if you can find a jar of mini cucumbers fermented in brine the old-fashioned way. Pickled sauerkraut has largely replaced real fermented sauerkraut as it is much easier to mass-produce and store long term. With recent interest in ferments, the tide is turning against industrial pickling and many people are prepared to seek out and pay more for fermented foods that have that extra health benefit.

Many fermenting microbes (yeast and bacteria) appear naturally floating in the air or on plant leaves and roots and can start wild fermentation. This is a form of fermentation where you literally start from scratch and create your own starter, attracting microbes floating in the air with foods like flour and sugar. Other microbes are bred commercially and deliberately added to the food as starter cultures in cultured fermentation (such as in yogurt, cheese or kefir). Milk-loving microbes often just need the right temperature to get started, whereas others really need a salty or sugary environment to succeed. Others may need to form a whole community to get going, such as in a kombucha SCOBY. For all of them, temperature and salt and sugar content, as well as the presence or absence of oxygen, are the key variables for successful fermenting that promote the key species above its other rivals, and importantly, keep the food safe. Where possible I have added some insights from our own recent analysis of commercial ferments: we looked at over a hundred products and detected more than 480 different microbes (although some were detected at very low levels). The analysis was a joint effort by my science team at ZOE, my lab at King's College London, who extracted the DNA, and my colleagues in Trento, who did the gene sequencing.

*

The chapters that follow are grouped into live and dead ferments. I start with the classic live ferments, where the microbes are still alive and can reproduce, to distinguish them from the not-so-live ferments, where the healthy microbes have been killed, inactivated or filtered out but may still have interesting properties.

## LIVE FERMENTS

# Yogurt

### How to make your own classic yogurt

- Gradually heat a litre of whole (ideally organic) milk in a pan to 85–89°C (or as close as you can without boiling or burning it).
- Leave at this temperature while stirring for 5–10 minutes to kill all the harmful bugs and break down the casein protein.
- Let it cool to 45°C.
- Add 2 tablespoons of a high-quality natural yogurt with no additives (starter).
- Decant into jars and keep them warm (if possible above 38°C) and untouched for about 8–12 hours. You can wrap them in towels and leave in an airing cupboard.
- Check they have set, keep a spoonful for the next batch and put in the fridge.
- It's that easy . . .

Fermented milk is simply a type of dairy product that has been cultured with specific bacteria or yeast strains, which break down lactose (the sugar in milk) into lactic acid. This process gives the milk a sour taste and changes its texture, making it thicker and creamier. There are many types of fermented milk products, including sour milk, kefir, buttermilk and labneh cheese, but yogurt is perhaps the most well-known around the world. It is made by adding specific bacterial cultures, such as *Lactobacillus bulgaricus* and *Streptococcus thermophilus*, to milk at a specific temperature. The bacteria ferment the lactose in the milk, producing lactic acid and giving the yogurt its tangy flavour and solid thick texture.

While the sensible way to make yogurt is from another yogurt (backslopping), there are other methods. You can also now make yogurt by picking your favourite microbes and buying commercial starters from a catalogue; breaking open some probiotic capsules that contain a mix of *Lactobacilli, S. thermophilus and Bifidus* to give you a healthy mix; buying some heirloom cultures online; or, if you are really brave, even try making your starter totally from scratch using milk then mixing in some ants' eggs, chilli stems or goats' droppings and waiting patiently for the LAB to appear. (Disclaimer: I have not tried this.)

I was brought up on SKI yogurt, a brand I associated with improbably happy and beautiful families dressed always in bright leotards bouncing on trampolines or skiing in the Alps. I remember them being quite sugary with chunks of fruit-like substances in them. Despite the ads, I never really considered them healthy, and certainly never thought you could make yogurt yourself.

Yogurt making machines have been around for a while but might have disguised how simple it is to make yourself without special equipment. Yogurt magically appears if you cool milk down after heating, allowing lactose-loving microbes to acidify the milk at just the right temperature and curdle the milk protein, producing that smooth semi-solid texture we love. The microbes involved are usually thermophilic, which means they are heat-loving and like to reproduce in moderate heat (43–46°C), although thinner and runnier yogurts can be made with hardier microbes that enjoy room temperatures called mesophilic bacteria. Most yogurts contain two to three microbes, usually a *Lactobacillus* and a *Streptococcus*, and generally come from added cultures or are 'backslopped' from adding a sample of a previous yogurt to milk. In the past, heirloom yogurts likely contained more species, and a few of these can still be bought today, though nobody knows how far back they go as the historical record of the microbes used is limited.

For thousands of years, yogurt has been a convenient way to extend the life of milk products. Although the word 'yogurt' is

Turkish, its precise origins have been lost. It seems likely that the first yogurt was created – probably by accident – when milk-producing animals were first domesticated in Central Asia around 7,000 years ago. Ancient versions also cropped up in Greece, India, Eastern Europe and elsewhere.

It wasn't until the early 1900s that interest in yogurt's health properties began. Russian scientist Ilya Metchnikoff noted that Bulgarian peasants, who regularly consumed yogurt, were relatively healthy and long-lived. He believed that many common diseases of the day were caused by putrefaction of the gut, and that the microbes in yogurt could reverse this. Convinced of his theory, he drank sour milk every day for the rest of his life, reaching the grand age of seventy-one – a good innings at the time. This was our earliest foray into probiotic therapy.

Metchnikoff met Louis Pasteur, who, having isolated lactic acid-producing microbes in 1857, was the first to prove that yogurt was alive. Today, we understand that certain microbes can benefit health, but back then they were only known to cause disease. Metchnikoff's work showed that the opposite could also be true, which pricked the ears of entrepreneurs, who saw a potential for profit. In 1919 Isaac Carasso founded Danone in Barcelona and created yogurt using two key microbes that Metchnikoff had isolated: *Lactobacillus delbrueckii bulgaricus* and *Streptococcus salivarius thermophilus*. Just a decade later, Dr Shirota founded Yakult, creating yogurt using a bacterium isolated from cheese, called *Lactobacillus paracasei shirota*. Both companies patented their strains of bacteria, upgrading yogurt from a peasant food to a pharmaceutical business and finally a multibillion-dollar food industry. This mass-produced yogurt was very different from traditional yogurt that had been made from simply adding a previous batch to warm milk and letting it cool slowly. The new methods of starter cultures made it easy to scale up industrially, but the resulting yogurt had lost the complexity of multiple strains in the homemade versions. They also found that while traditional cultures lasted indefinitely, the new lab-grown culture varieties would only produce a few batches before needing renewal. The reason is probably that the traditional cultures have had millions of generations to evolve into species that cooperate to survive. Having more different species working together also meant they wouldn't all be killed by the same predators, like rival bacteria or virus phages. These can destroy a commercial batch if infected.

The simple fruit-flavoured yogurts I ate as a child are mostly now replaced by hugely profitable ultra-processed products, often aimed at children. These are clearly not healthy, but what about so-called natural yogurts?

## Yogurt and health

This question has been debated since Metchnikoff began working with Pasteur and realised the potential of microbes for health. The market is dominated by some of the world's wealthiest companies, often in the past reprimanded for over-hyped health claims and advertising. Cultured microbes in yogurt really didn't hang around and multiply in the gut to the huge extent that adverts of ten years ago suggested. A 2017 meta-analysis from nearly 300,000 people showed that eating yogurt episodically did *not* reduce heart disease risk. However, another review showed consuming at least 200g per day (over half a small pot) was associated with a lower risk of cardiovascular disease. In 2022, a larger review including data from 896,871 participants found that mortality overall decreased by about 7 per cent, and mortality from heart disease by 11 per cent for those consuming yogurt, with the most benefit seen for those who ate the most. Interestingly, there were no clear benefits found for cancer.

As these studies were observational, it may have been biased by better lifestyle and other habits associated with eating yogurt, although if true, we might expect reductions in cancer, too. Fortunately, several small randomised clinical trials (with less possible bias) back up these results and suggest yogurt may also improve levels of blood sugar and blood fats.

If yogurt supports heart health, can it also support a healthy weight? There are only two small randomised clinical trials investigating yogurt and weight loss, and both were inconclusive. However, nine of ten observational studies that followed more than 219,000 people eating yogurt showed weight loss or a reduced waist size compared with consuming milk.

In 2018 my team looked at the yogurt-eating habits of nearly 2,000 UK twins and their gut microbes using the latest microbiome sequencing methods. The microbiome diversity of the regular yogurt eaters was significantly greater (i.e. healthier) than non-eaters, and we consistently (and reassuringly) found higher levels of the culture microbes *S. thermophilus* and *B. lactis* in these people. The rates were highest in regular eaters and lowest in occasional eaters, confirming that the microbes don't hang around for long. We also found in the blood and stools of regular yogurt eaters many chemical metabolites (produced by microbes) that are likely protective against developing internal belly fat. A more recent Japanese clinical trial of 100 overweight men gave them placebo or yogurt fortified with *Lactobacillus plantarum* and found that after twelve weeks they had significantly reduced their abdominal fat. So we think that the probiotic microbes in yogurt either have a direct effect via the chemicals they produce, or indirectly stimulate our gut microbes to produce healthy anti-inflammatory metabolites that help regulate our weight and metabolism. This may be an example of the new era of the 'postbiotic' effect as we learn to harness the healthy chemicals that our microbes produce.

Yogurt could also help the immune system and certainly seems to help pigs and mice fight a range of respiratory and gut viruses. In humans, there are a few, less convincing studies, such as one where volunteers ate either a large pot of yogurt or a control for nine weeks, and the yogurt eaters saw improved markers of inflammation and immunity. Our own unpublished twin studies show clear differences in many immune markers in blood between yogurt eaters and non-eaters, although we don't know their clinical significance yet. Our Covid studies also showed people taking probiotics were 14 per cent less likely to get infected, so we might expect yogurt and other fermented milks to have a similar effect. More recent studies have been clearer, with sixty-four Japanese students eating either placebo or yogurt containing *L. bulgaricus* for twelve weeks. This showed that yogurt eaters had a better response to the flu vaccine and led to fewer infections. A little word of caution comes from studies of heavy yogurt consumers in China, which show changes that could accelerate antibiotic resistance; the theory being that the process of gene transfer to nasty pathogenic bugs is accelerated. The way to mitigate this is to ensure that yogurt consumption is part of a healthy gut-friendly and plant-diverse diet.

## What to look for when buying yogurt

First, always read the label – not all yogurts are equal, and the differences between products can be vast. It can be difficult to find full-fat yogurts, and the majority contain added sugar and flavourings, neither of which will benefit your gut microbiome. Sugar prevents microbial growth, which is why jam and honey have such a long shelf life, while flavourings, even less frightening ones like vanilla, tend to be synthetic, and we have no idea how they impact gut bacteria.

As yogurts shifted from peasant fodder to the modern mainstream, food manufacturers jumped wholeheartedly onto the bandwagon. To maximise returns, they now market their brightly coloured, sugar- and chemical-filled yogurts to children. A survey of 900 yogurt brands in 2018 found that 90 per cent of products were high in sugar; this figure was even higher for children's yogurts. Food companies are hoping that these fake yogurts might help fill the void created by the crumbling market for children's sugar-filled cereals. Despite the smiley faces and pictures of fruit and vitamins on the label, they should be avoided – the sugar these products contain accounts for at least 45 per cent of the energy per serving. That this is still allowed is a disgrace. Our children deserve better protection. We urgently need legislation and clearer labelling to help parents see past the disingenuous packing and health claims on these products. The ban on advertising these and similar products to children in the UK is a good first step, but we need to go much, much further.

Low-fat yogurts are still incredibly popular (particularly in young people), despite zero evidence that they are in any way beneficial to health. Without yogurt's natural fat content, it has a much less pleasant taste and mouthfeel. To counteract this, in the vast majority of cases, companies add ultra-processed food derivatives, including modified tapioca, maize starches, emulsifiers, stabilisers and guar gums. It is also common to add sugar, fruit extracts and other flavourings to distract the consumer from the lack of delicious fat. Although all yogurts contain live cultures initially, without whole milk and surrounded by large amounts of sterilising sugar, those beneficial microbes are unlikely to survive. However, our survey found that all yogurts tested (even the sweet ones) had detectable microbes – mainly *S. thermophilus*, and one or two others in small amounts.

There's no evidence that low-fat natural yogurts are any healthier than the full-fat options, although some of the microbes present in low-fat versions may still hold health benefits. When manufacturers remove the fat from whole milk to make skimmed milk, many of the fat-loving vitamins are taken out with the fat. This means that low-fat yogurts produced with skimmed milk have fewer of these vitamins, which manufactures often add back in. Without any health benefits associated with low-fat yogurts, this seems to be a monumental waste of time and energy. As long as your yogurt is natural and alive, there is no need to choose low-fat – unless you prefer the lack of taste.

Natural yogurts on the other hand, generally have a higher percentage of protein, more live cultures and significantly less sugar than their highly processed counterparts. Let's consider some of the common options available.

### *Greek yogurt*

While most Greek yogurt (and the milk it was made from) has never been anywhere near Greece, the name is now synonymous with thick, full-fat, strained and healthy yogurt. As the whey is removed, most of the lactose is also removed. In the UK, it tends to be called

'Greek style', while elsewhere you can find similar products called Turkish yogurt or labneh in the Middle East. Having Greek or Greek style on the label does not necessarily guarantee you a pure yogurt, though. Often, to save pennies, companies will add thickeners, like pectins and starches, and even artificial powdered milk.

### *Icelandic Skyr*

This is another type of yogurt that is gaining ground on Greek style yogurts, with a lower fat content and more protein while maintaining the desirable mouthfeel. Skyr, however, is quite a different product; it is not technically a yogurt as it is not strained and often contains rennet, used in cheese production. It usually contains the cultures *S. thermophilus* and *L. delbrueckii*, as in yogurt, but unlike yogurts yeast is often added. It is more expensive and uses four cups of whole milk for every one cup of yogurt, making it less good for the planet.

### *Fromage frais*

Another mainstay in UK supermarkets, mostly aimed at children, which, as you might have already guessed, means it is generally high in sugar – often 3 teaspoons in each tiny pot. Some manufacturers even add children's sweets containing 19 per cent sugar. Fromage frais is not, in fact, yogurt, but a young, unripened fermented cheese, which means it is lower in fat and, according to French law, must contain live bacteria. However, due to the high sugar content, these microbes cannot thrive. A similar product, fromage blanc, is not required to have live cultures, as fermentation is halted early.

### *Non-dairy yogurt*

Many people assume (and I was one of them) that using non-dairy milk substitutes in yogurts would naturally be healthier, because of the lower saturated fats. But in general commercial varieties are slightly less healthy, either because they contain many chemicals to reproduce the dairy (emulsifiers and thickeners) or they use sugary starches to add solidity to the product. In general they do have slightly more protein and less fat, but way more carbohydrates (sugars). A 2021 US study examined 249 non-dairy yogurt alternatives. At least one-third had good (5g or more) of protein per serving, but less than half had adequate calcium levels, while only about one in five had adequate vitamin D and B12 fortification. One half of the dairy alternatives had high sugar levels, so are best avoided.

I was surprised to learn that the microbial composition of plant-based yogurts (such as oats) is fairly similar to milk versions. Some use the same *Lactobacillus* and *S. thermophilus* strains as milk-based ones, but others rely on microbes that eat whey protein or the specific plant used, such as coconut, oat, soy or nuts. There is little data comparing the microbial efficacy of the products, though one review suggested that microbes might not survive as well in some non-dairy products. In conclusion, we don't yet have enough data to really compare them, but if you are vegan or lactose-intolerant then plant-based yogurts appear a good substitute. But try to avoid ultra-processed products with long lists of hard-to-pronounce ingredients, or those containing added sugars.

**Coconut yogurt** seems to be the best non-dairy option. It is made with coconut fruit flesh, which has less fat content than coconut oil and importantly plenty of fibre, and by avoiding dairy is much more environmentally friendly. It certainly tastes good and is simple to make without too many additives, but does it confer the health benefits of traditional 'live' yogurt? It seems that it could do, at least in terms of probiotic content. It doesn't compete with cow's milk yogurt protein content, as it only serves around 2g per 100g compared to natural set yogurt at 10g per 100g, though fortification can boost calcium. A company creating coconut milk kefir claims to have created the most powerful probiotic currently on the market, with over 4 trillion CFU and 40+ different types of probiotic strains in just 1 daily tablespoon.

### *Yogurt in five*

1. Yogurt usually contains around three bacteria (although mainly *S. thermophilus*) and is a healthy food especially when compared to milk.
2. Reasonable evidence suggests that it reduces heart disease risk, inflammation and obesity.
3. Fat free versions or those with excess sugars, additives and sweeteners are unhealthy.
4. Non-dairy versions, if low in additives, are likely to also be healthy.
5. Both dairy and plant yogurts are easy to make cheaply at home.

## Kefir

### How to make your own aerobic kefir

- Start with a clean 500ml glass jar with a cloth or paper cover.
- Add 1–2 teaspoons of kefir grains.
- Add whole milk (preferably organic but can be any type) and put a cloth over the top (this allows for anaerobic fermentation).
- Leave on a worktop but out of sunlight for 18–36 hours to ferment. Stir or shake occasionally.
- Taste and sniff regularly to work out when it's producing acid and you are happy with the flavour.
- Strain the mixture into a clean container, separating out and keeping both the grains and the kefir drink separately in the fridge.

Meaning 'good feeling' in Turkish, kefir is another form of fermented milk, but it is more liquid than yogurt. Historians believe it first originated in the Caucasus mountains, which intersect Europe and Asia. Wherever its precise origin, we know it has an ancient pedigree. Kefir was found in an amulet around a 4,000-year-old mummy's neck and has since been traced to 3800 BCE in early Bronze Age China. Thankfully, things are changing fast and kefir can now be found in the refrigerated aisles of many supermarkets, at least in most urban areas of the UK, Europe and the US.

Around the rest of the world, kefir is enjoyed in many different variations and comes in a range of forms depending on the base, including milk kefir, water kefir, and coconut milk kefir. The principles are similar but methods of making (and eating or drinking) them vary. In Russia, it is often consumed plain or used in soups and stews, while in the Middle East it is commonly served with honey and nuts. In Central Asian countries such as Kazakhstan, where I visited a few years ago, it is often made with mare's milk (definitely an acquired taste), in the Pyrenees with sheep's milk, while in the United States it is typically made with cow's milk. Overall, kefir is a versatile and nutritious beverage that has a rich history and global appeal.

Milk kefirs are a type of dairy product that have been cultured with specific bacteria and yeast strains, or kefir grains, all of which break down lactose (the sugar in milk) into lactic acid. This process gives the milk a sour taste and changes its texture, making it thicker and creamier. The unique natural properties of the kefir grains and the microbes that they contain make kefir distinct from other fermented milks and results in a greater complexity of flavour. Kefir is very simple, cheap and quick to make.

Kefir can be made from any dairy milk and many plant milks. Kefirs using soya or coconut milk have a nice creamy consistency; other plant milks are likely to make a thinner kefir. No one knows where the original kefir grains came from that have produced the offspring we use today, but there are some interesting theories. According to some legends, the Karachai people of the North Caucasus received them as a gift from the Prophet Muhammad himself. Kefir grains, the necessary starting point for all kefirs, are a type of SCOBY (symbiotic culture of bacteria and yeast) and look like a little collection of jelly-like cauliflower heads thanks to their textured creamy appearance. Whether making goat's milk kefir or plant-based milk kefir, the grains are essential, but it's the type of milk you choose that mostly decides its flavour and texture. Goat's milk kefir, for example, has a thinner texture than cow's milk. Plant milks tend to make a fizzier kefir due to their sugar content. The kefir grains are strained out of kefir before it's consumed and can be reused many times (much like the SCOBY for kombucha, as we'll see later), making kefir cheap and the grains a very good investment. Kefir grains are stored in the fridge in a grain hotel (a large jar with a lid that contains enough milk to keep them alive for a few weeks) until the milk is changed or the grains regenerated. A similar principle is used for tibicos (water kefir) using sugar and water instead of milk.

But what are these cauliflower-headed grains? Containing up to twenty times more varieties of microbe than yogurt, they are the dried remains of old kefir. When we tested commercial kefir grains, they usually had greater numbers of species than commercial kefir milk products. They house a complex bacterial society, held in suspended animation, ready to get to work as soon as their environment suits their needs. They include the usual milk-loving, acid-forming lactose bacteria (*Lactobacilli, Lactococci*), plus the acetic acid formers (*Acetobacter*) and some carbon dioxide-producing yeast like *Saccharomyces* to give it fizz and small amounts of alcohol that can, in some rare cases, reach 3 per cent. No two kefirs will taste or smell exactly the same. We sequenced commercial kefirs and found a wide range of microbe species, between eight and thirty, and some older reports record up to sixty-one species for traditional kefirs that may be underestimates. These numbers vary widely depending on the milk and environment, and how long the fermentation takes.

The biology and true origin story of kefir grains is an incredible example of the power of evolution and symbiosis. At some point in time, thirty to sixty species of microbes, varying enormously in size and function, decided to pool their resources and work as a team. By coming together they have, in fact, become immortal. Kept in the right environment and fed the right food, they live for ever. Unfortunately, their immortality needs diligent caretaking. I have neglected my grains on numerous occasions and inadvertently killed them. To check your grains are still alive, try to make a new batch. If the resulting drink is sweet rather than sour, and there are none of the usual signs of fermentation like bubbling, they are sadly deceased. Mould or strange smells are another telltale sign that they have perished.

Scientists have devoted lifetimes to studying these fascinating grains, which should, perhaps, be renamed kefir *bodies* as they act as living organisms, splitting off babies, reforming, and sometimes combining with other microorganisms if advantageous. Try though they may, scientists have still not been able to recreate this natural wonder in labs. Adding the thirty or so species together, and hoping they get along, does not a kefir grain make.

Kefir varies by region – Tibetan kefir apparently has the most microbe species (measured at over sixty), although it is a bit of a trek for most of us to get regular supplies. Luckily pretty much any kefir grains will work well. Their microbes contribute hundreds of different complex chemicals that influence flavour and health properties.

The true magic of kefir grains comes from their ability to work as a team. The microbes within them manufacture simple sugars – glucose and galactose. From these basic building blocks, they create an exopolysaccharide (EPS) called kefiran, which forms a protective shield around the microbial community. This allows them to dominate their milky universe, ensuring that would-be microbial invaders cannot get a foothold. Some researchers believe kefiran itself may help support our immune health, although studies to date have been small. Either way, kefiran helps give kefir its thick texture. Even stranger is the fact that the microbes in kefiran are not quite the same as those found in kefir milk. This suggests that the kefiran is not just acting as a concentrated powder that dissolves into the milk but is, in fact, controlling the microbes in its surrounding environment. We still have much to learn about kefir.

## Kefir and health

Kefir can have as much as ten times more probiotic (healthy microbe) content compared to yogurt, so it promotes digestive health and immune function to an even greater extent. Also, the protective structure of kefiran likely helps the microbes survive longer, even in the harsh environment of our gastrointestinal system. It is a good source of calcium, amino acids, B and K vitamins, and vitamin D. Just one cup provides up to 10g of protein and 30 per cent of your required vitamin B12. One ZOE member, Frank, received his ZOE retest results after six months and found he had improved his ratio of healthy microbes. He told me with great pride that he'd achieved a very high ZOE microbiome score, and that this was because he had started having kefir every morning to help him and his microbes recover from an illness. In Frank's case, a daily breakfast of kefir had clearly worked. He had managed to boost helpful strains and reduce unhealthy ones in his gut. But that's just one anecdote – does the science back this up?

While there are many positive kefir studies, most are quite small, usually carried out in a test tube or in rodents. But they are consistent in showing a number of beneficial effects, including reductions in blood cholesterol, blood sugar, blood pressure, weight and inflammation. There is also indirect evidence based on reduction in immune markers of a potential anti-cancer effect. Several studies have consistently shown a positive effect on depression-like symptoms in mice, with improvements to serotonin levels and gut microbiome composition. In addition, there are now a few decent clinical studies in humans that report benefits, although results are mixed. One study in patients with type 2 diabetes failed to show any blood fat reduction, but others have shown benefits, mainly and most consistently on lowering levels of blood glucose. Given the diverse microbial make-up of kefir, this makes sense.

Indirect evidence comes from a systematic review of multiple studies in 2018, which found that the same probiotic microbes found in kefir help (on average) to improve symptoms of irritable bowel syndrome (IBS). Another clinical study in twenty-eight overweight subjects showed benefits of kefir on protecting the gut lining, which is likely to be important in IBS. There is also some preliminary evidence to suggest that kefir may have beneficial effects on mental health. Kefir consumption is associated with reduced symptoms of depression and anxiety in some smaller studies, which is promising.

So, while we still need much bigger human studies and don't yet really understand the mechanisms because of the complexity of kefir itself, there's little doubt it works. The evidence so far of the multiple small studies is reasonably convincing, but if we combine it with the impressive Stanford clinical trial of fermented foods it gets even better. This 2022 study showed significant reductions in blood inflammation and improvement to the immune system in just four weeks of eighteen people taking five portions of fermented foods daily, of which the commonest was kefir. While the results in mice and rats are clearer than in humans, there is currently good evidence of an anti-inflammatory and anti-blood sugar effect of kefir in humans, in addition to its benefits on our gut microbes. Kefir may well have other benefits, such as improving our mood and sleep. Although we couldn't easily separate out high kefir users in the 2024 ZOE fermentation study, people who had an average of four portions of kefir a week improved their mood and energy scores significantly compared to baseline.

## Making kefir at home

Living kefir grains have minds of their own. Sam and Damien make their handmade Ki Kefir next door to an excellent organic, grass-fed dairy farm in East Sussex. For a long time, their goal of making a consistent product seemed impossible. This is because kefir grains are incredibly responsive to everything in their environment – right down to the microbes that you, as a fermenter, pass on to them. The grains respond to the type of milk they're added to, how frequently you stir them, the air temperature, and how many yeasts they contain, determining how fizzy and frothy the final product is. Embracing inconsistency is key with kefir; it's living food and there's only so much you can do to make it behave the same way every time. It's easier to achieve something close to consistency in winter, but the milk itself changes throughout the seasons with the cows' diet. In summer, the cows eat more grass, making the milk fattier and giving it a fuller flavour profile, while in the winter the lower fat percentage results in a fizzier kefir.

To make kefir at home you simply add kefir grains to milk and leave it to ferment at room temperature. During this time, the bacteria and yeast consume the lactose in the milk and produce lactic acid, which gives kefir its tangy flavour, as well as creating carbon dioxide, which produces the effervescence and hint of alcohol. Kefir is wonderfully simple to make. I ordered some kefir grains online, added a sachet to a litre of whole milk in a jar and left it in a warm room for thirty-six hours until its smell and a few bubbles told me it had turned acidic. I then sieved the lumpy grains out and put it back in the fridge in jars with a drop of milk for the microbes to continue feeding on and keep them alive. The old kefir contains starter microbes for a new batch so, like sourdough starters, if they're well fed you can keep them alive for around six batches, making this a very cost-effective method.

If you don't have grains or a handy friend from whom to beg, borrow or steal, my favourite hack is to buy the best commercial kefir brand you can find – free of flavourings or additives, ideally with a bit of fizz and that has not been killed off via pasteurisation – and use this as a starter. Simply add a large tablespoon of the bottle to a warm pint of milk and wait around thirty-six hours to transform (or shorter if the air temperature is warm). You'll need to strain the fermented batch with a conventional sieve or through a tea towel to separate out the lumpy grains for reuse before putting the kefir in the fridge.

Although the process is simple, don't expect the same results every time. My efforts always end up tasting slightly different, and that's part of the charm. Through trial and error, you can slowly unpick which factors make the greatest difference to the final product. By trying different grains, milks and temperatures, you can learn how each influences the flavour and fine-tune it to your liking. With this many microbes and different conditions, they do vary a lot.

As with yogurt, coconut and soy milk are a good alternative to cow's milk because of their protein and fat content. A tin of full-fat coconut milk and 2–3 tablespoons of kefir grains, left to ferment for three days, will make a tasty and live coconut kefir.

## Commercial kefirs

You can now find kefir in supermarket aisles across the globe. Brands in shops can vary widely, but those that are very mild, thin and consistent are now not usually made with kefir grains, nor are ones that have longer shelf lives or nationwide distribution. The unpredictability of using kefir grains is a real problem for large companies. For this reason, most have pragmatically switched to use what they call 'kefir cultures', adding commercial cultures of many (between eight and fourteen is routine) of the common microbes found in kefir grains. They have less diversity of microbes than homemade or traditional versions and many of these lack the carbon dioxide-producing yeasts that provide the fizz and froth. But on the plus side you are getting around a dozen different strains, four times more than most yogurts, and they are less likely to explode or go sour rapidly.

So, not all kefirs are equal. Studies have shown that traditional kefirs made from grains have a much greater effect (35–55 per cent) on reducing blood fats and weight (at least in mice) than commercial ones made with a more limited set of microbes with the yeasts. But it is not all bad news for commercial kefirs. One study compared the microbes in a traditional versus commercial kefir and reassuringly found a similar number of microbe species, although the traditional brews had a stronger potential action against other infectious microbes. I have seen some kefirs that are pasteurised, meaning that all the microbes have died. I would have dismissed these as useless but surprisingly, these may still be beneficial. A group led by my academic colleague from Cork, Paul Cotter, actually found clear benefits in mice when fed kefir grains that were heat-treated and no longer living.

Browsing the aisles and online supermarkets, you'll find at least five types of kefir, ranging from the worst (a category that should not even be called a kefir) to ones that are made according to traditional methods.

The worst category I found is a 'kefir' aimed at children, which doesn't even contain what would be considered yogurt. Made with milk powder, cream and standardised milk, as well as added fruit sugars and emulsifiers, they had added some commercial live cultures to the finished product in order to give this unhealthy ultra-processed food the 'kefir' halo name. So beware of being conned.

The second type of commercial kefir is also ultra-processed and is more of a supplemented super-yogurt flavoured with fruit juices or other additives and sweeteners. While having these occasionally is not a problem, I would not consider it a healthy food to eat every day. Given the choice I never buy flavoured 'kefirs' or those with extra fruit, added sugars or artificial sweeteners, as you never know what effect the additives have on the microbes. Artificial sweeteners all have some disruptive effect on our gut microbes, so it is always better to add your own fruit and sweeten it yourself.

The third, and most common, are essentially yogurts enhanced with the probiotic strains normally found in kefir – let's call them kefir culture yogurts. They generally do not contain *Acetobacter*, the kefir-specific *Lactobacilli* that live off the grains (e.g. *L. kefiri* and *L. kefiranofaciens)*, or any yeast species, all of which are frequently found in traditional milk kefir made from grains. These subtle differences may be important for health benefits. For example, *L. kefiranofaciens, L. kefiri and Kluyveromyces marxianus*, which are often absent from commercial kefirs, have individually been shown to promote improved mucosal immunity and metabolic health, and protect against colitis in animal models. Our survey found that the cheapest supermarket kefirs from Poland had only six species of bacteria.

A 2023 Canadian-Irish study explored two types of cultured kefir in twenty-one healthy male volunteers in a pilot study. One kefir, the most commonly available commercial brand in Canada, was made with 1 per cent fat milk and contained more than twelve kefir microbes but lacked *Acetobacter* or yeasts. The other was a special concoction of nine microbes, many of which had been associated with traditional kefir and with benefits in mice. They found the tailored concoction they had created (and patented) had greater reduction of blood fats and lower potential cardiac risk. Single-strain drinks such as Yakult have been popular for years. We confirmed the only microbe present was *L. casei Shirota*.

Cultured kefirs with multiple strains are here to stay and are consistently better than yogurts that typically have three or four types of microbes where these super-yogurt cultured kefirs will have twelve or so. Cultured kefirs are shelf stable and tend to be similar in taste every time you buy them. When we designed our supermarket ZOE gut shot, we used a fourteen-microbe cultured milk kefir as our base, as using grains was just not practical for a mass market needing shipping and storage across the country. Using grains and more microbe species would mean less consistency, not to mention produce gas that could cause minor explosions. This is a new field and adding more healthy strains will produce even healthier products.

In California, you can now buy an actual raw milk cultured kefir (from Raw Farm), which has the benefit of extra bugs in the raw milk (it gets tested for pathogens before being sold) but can't travel far. Raw milk is growing increasingly popular, but I personally advise against drinking it, unless it is transformed into traditional cheeses with good safety records. The health claims associated with raw milk are not based on scientific evidence, and, nutritionally speaking, pasteurised milk is almost identical to raw milk. Most worryingly, raw milk is unpredictable and can harbour disease-causing microbes like *Listeria* and *Salmonella*. Unpasteurised dairy products (mostly milk and homemade cheese) cause 840 times more illnesses than pasteurised dairy products in the US, and are forty-five times more likely to put you in hospital.

The fourth category is kefirs that are made at a smaller scale using real kefir grains in whole milk, usually from goats, cows, sheep, soya or coconut. These include brands such as Ki Kefir or Chuckling Goat in the UK and contain the yeasts as well as multiple microbes, creating a complex flavour profile and fizziness that changes unpredictably from day to day as the fermentation process continues. They have to be drunk within a short time scale to avoid accidents. The Cultured Coconut brand from Canada has made 'medicinal' coconut kefir with over 40 strains and 4 trillion colony-forming units (CFU) per tablespoon. We tested two coconut milk kefirs in the UK and while not quite as impressive they still contained between eight and sixteen species of microbe.

The fifth type is one that you would make at home using kefir grains in your choice of milk, where you can experiment with flavour and fizziness. This type is the cheapest of the lot and packs by far the most probiotic strains, complex nutrients, health benefits and personalised flavour. Commercial grains for home use bought online are often dried powders that can contain both dried grains and a selection of six or so culture bacteria and yeast, but vary widely and are not well monitored, so go on recommendations. I have tried several and they all seem to work and allow me to produce grains of my own. Because of work and travel commitments, I have to be flexible, so I have found this my preferred choice; I find the smell and fizz strangely comforting and the ability to 'backslop' and top up my supplies with fresh milk reassuring. When I don't get my daily shot, I feel like something in my life is definitely missing.

## How to eat kefir

Purists may enjoy their kefir as a daily shot in a small glass, but don't stop there. There are many other simple uses for it, which are especially useful to have in mind when you have made a big batch.

- Add as a side to a curry or chilli with a squeeze of lime or lemon.
- Try adding them to your smoothies in place of (or in addition to) yogurt or milk.
- Make ice cream by blitzing in a blender with frozen mango or frozen bananas and a dollop of nut butter.
- Try mango lassi – an Indian favourite if you want a sweet taste. Simply mix some mango pulp in a ratio of 1:5 with kefir.
- Kefir salad dressing is a favourite at my home: mix kefir, extra virgin olive oil, a squeeze of lemon or lime, garlic, green herbs (parsley, dill, chives, mint or coriander), salt and pepper and blend – add half an avocado if you like a richer texture. Add to a chickpea salad for an instant flavour and health boost.
- Make your own labneh. Strain your kefir through a paper towel or coffee filter overnight in a cold room or fridge to create a kefir cream cheese, or coconut cream equivalent, and flavour with herbs and spices. You can eat it like soft cheese or roll into little balls and store in olive oil.
- Add to dishes or at the end of cooking instead of cream or crème fraîche, for example in a mushroom stroganoff or a veggie curry, or blend with avocados in a chickpea sandwich (see *The Food for Life Cookbook*).

To change the flavours and fizziness in your kefir, try adding slightly more grains or changing the milk you use for your base, or you can leave it out at room temperature for slightly longer for a more sour taste and more fizz. After straining your kefir grains, you could add some nutmeg or cinnamon before serving, blend it with fresh fruit, or serve it with nuts and honey.

## Kefir troubleshooting

- Knowing when it's ready will depend on conditions, but smell is a good guide – if it smells rotten or foul, that's a bad sign, as is seeing the grains float to the top of the curd.
- Pockets of lumpy whey can form, especially in warm weather or if left too long. This is not a problem and they can be reincorporated with stirring or whisking.
- Kefir can curdle into lumps, particularly in summer. It will appear lumpy and lose its distinctive sour smell. It means other spoilage microbes have taken over and, unfortunately, you will need to abandon the kefir and retry in a cooler space.
- White film can start growing on top. This is rare, but may be kahm yeast, which is not dangerous but can mess up a batch. Rescue the grains and put them into a new jar with no air.

I am addicted to my daily dose of kefir and the process of making it myself. As well as my morning shot, I now add kefir to many spicy dishes just before serving to enrich the sauce or cool my palate. The evidence is overwhelming that consuming these thirty odd microbes regularly is good for you, but importantly also very tasty. Let's do our bit to keep these immortal grains alive.

### *Kefir in five*

1. Kefir has been part of food culture across the globe for centuries and can be made from fermenting cow, goat, sheep, horse, or camel milk, as well as soya and coconut milk with starter cultures.
2. Having kefir every day can help maintain a healthy, diverse gut microbiome and support immune system function.
3. When buying kefir in a shop, look out for local brands with short shelf lives that use kefir grains; otherwise pick cultured kefirs but avoid artificially flavoured and sweetened ones.
4. Use my simple recipe (see page 242) to make your own at home and enjoy it with fresh chopped fruit or as a dressing for salads or a creamy addition to soups and curries.
5. Kefir made from grains is a rich source of around thirty healthy probiotics as well as amino acids and bioactive compounds – an easy and cheap way to enjoy a very high protein drink.

## Tibicos (water kefir)

### How to make lemon, ginger and turmeric tibicos

- Add 2 tablespoons of tibicos grains and 50g sugar to a litre of water (tap water is fine, preferably filtered).
- Ferment for 24 hours in a closed jar.
- Gently strain out the grains and place them back in the fridge.
- Add half a sliced lemon, an inch of ginger and a teaspoon of ground turmeric.
- Continue fermenting (lid closed) without the grains for a further 24–36 hours.
- Decant into closed-lid bottles and leave out for another day to get more fizz.
- Chill and store in the fridge and then enjoy.

Despite its name, water kefir is completely different to dairy kefir and is more like kombucha. I think a better name to use is the Mexican tibicos, but it has many other names including Japanese water crystals and, strangely, California bees. I only discovered it properly recently and, for me, if you like a bit of fizz, it comes as close to the flavour of alcohol as any soft drink. It is made from mixing specific water kefir grains with sugar and water to get the basic brew in a day or two. You can then add fruits or herbs afterwards to change the flavours and add fizz. It is milder and less acidic and tart than the better known kombucha, making it more flexible flavour-wise. While its precise origins are unknown, it was well known in the Caucasus mountains, Tibet and Ukraine. It may have originated from the unlikely sounding (and made-up) ginger beer plant and was brought back to Britain by soldiers fighting in the Crimean War. There is also documentation from Mexico in the 1890s that tibicos grains were used to make a soft form of tequila from the juice of a prickly pear cactus, where the grains naturally formed.

The advantages of tibicos are that you can make it very fast, usually within two days, and because the microbes rapidly consume it, the final sugar concentration is lower than in other ferments and it is only mildly acidic. What milk and water kefirs share is an essential group of symbiotic microorganisms that form the kefir grains and the complex exopolysaccharide (EPS) sugar backbone, but rather than only feeding off milk, water kefir grains feed off any form of sugary water and are much more versatile, although the grains are a bit more sensitive and delicate than milk kefir grains or kombucha SCOBYs.

This versatility of the microbes on the grains means they also feed off most sugary fruits, some vegetables, many nut milks and coconut milks. The grains are more translucent and jelly-like than those used for dairy kefir and you can buy them online or get them from a friend. But be careful as they are sensitive creatures: they don't like being handled or picked up with hard metal and – as I know from personal experience – they do not like being abandoned. I was given some grains by a Belgian friend with instructions about feeding, and I wrongly assumed they would be indestructible, like my kombucha SCOBY. Generally the grains prefer something in addition to the sugar and water to thrive, and there are many options and recommendations from experts. These include adding ginger, sea salt, baking soda, spices like turmeric or small pieces of fruit like figs. Figs are often used to combat any hard water in your area, which can increase calcium and phosphate levels.

I made my first fruit kefir (pomegranate fizz) with Instagram star Julius Fiedler. As we filmed together and I talked about microbes, he talked me through the process, which is actually much easier and quicker than it sounds, although you do have to be well organised. The big advantage of water kefir is that you can be in charge of whether you want more or less fizz, more or less acidity or more or less sweetness or fruity tastes. It can be made in a closed (anaerobic) environment, which I prefer, or an open (aerobic) one. I prefer more fizz and slightly less acidity and alcohol and keeping out oxygen favours the LAB rather than the yeasts and *Acetobacters*.

The many microbes in tibicos resemble those in kombucha but with more LAB. They are actually closer in composition to milk kefirs but with marked differences in the types of LAB and yeasts, as well as some *Acetobacters*. Hundreds of different microbes have been recorded, but the main ones common to most brews are *L. paracasei*, *L. hilgardii*, *L. nagelii* and the brewer's yeast *Saccharomyces cerevisiae*. Some studies suggest even more variety, with *L. ruminis* being the commonest bacteria in all samples and *Pichia* the commonest yeast out of over ten commonly found in all samples. As a relatively unknown newcomer, *L. ruminis* is a likely human probiotic bacteria; as the name suggests, it is also found in ruminants like cattle and horses and has been shown to have strong immunoprotective and antiviral effects in laboratory studies.

A yeast called *Dekkera bruxellensis*, known for contaminating Belgian beer, is also commonly found, which is one reason brewers don't like fermenting in the same areas as kombucha or kefirs as microbes can easily hop over and ruin a brew. Providing a mild pleasant tang, acetic acid bacteria are present in low levels unless there is plenty of air in the fermentation; interestingly, these bugs are mainly found in the liquid not the grains. As with its dairy-loving cousin, tibicos produces a similar sugar-based dextran EPS network that the microbes hang on to for protection. The *L. hilgardii* and *L. nagelii* microbes are like specialist bricklayers: when they find suitable sugar molecules they break them up and reassemble them into more of the EPS structure. This is how the tibicos grains can build up rapidly with sugar. Only a few makers of commercial tibicos have tested the microbes in their product. One exception is a water kefir made with coconut (called The Cultured Coconut, produced originally in Canada) that claims to contain over a trillion CFUs and around forty microbial species. It's made with tibicos grains, chopped coconut flesh and no added sugar. (Note this is confusingly different to coconut milk kefir, which is a vegan alternative to dairy and uses milk kefir grains plus coconut flesh – see page 69.) We tested some water kefirs which, surprisingly, performed very well. The grains had a greater number of microbe species than milk kefir, and Agua De Madre Elderflower & Apple, which comes in a bottle with a short shelf life, had a count of more than eighty species. All the bottled water kefirs we tested contained yeasts. We also tested a can of water kefir; the number of microbes dropped to a third.

The rapid fermentation process can be slowed down by reducing the sugar, reducing the temperature or adding fewer grains to start with. Some experts recommend removing the grains after twenty-four hours to protect them from exhaustion, even if the brew is not fully fermented, and then allowing it to continue at a slower pace for another day. This worked well for me in the heat of summer. The liquor surrounding the grains has a similar composition in terms of microbes and chemicals so when the liquor alone is used as a starter, it is slower to ferment but can be an alternative to using the fragile grains if you are not in a rush and want a slower ferment.

## Tibicos and health

There are only a handful of studies looking at the health benefits of tibicos, two of which showed benefits for bones, blood fats and weight – in rats. In humans, I haven't found any clinical studies published, but one lab study looked at the effects of water kefir on an artificial gut that simulates the environment and sections of the human intestines. They observed an increase in the production of short chain fatty acids, which are good for the gut lining, increased levels of helpful *Bifidobacteria* microbes and a general improvement in the gut barrier to stop leakage. What was really fascinating was that the results were slightly better when the tibicos was pasteurised and so effectively dead. Again we see evidence that the post-biotic chemicals that microbes produce work well after death, which adds value by extending the shelf life and stability of the product. Despite the disappointing lack of human data, the sheer number of probiotic species present and the hundreds of chemicals they produce suggests drinking tibicos regularly is very likely to be good for gut health. The modest amounts of natural sugar of 1–2 teaspoons per bottle or can shouldn't be a problem for most people, compared to other sodas that have four times as much or most alcoholic drinks.

## What to look for when buying commercial tibicos (or water kefirs)

I love making my own tibicos as it is so versatile and quick to make. Commercial brands do exist but they are still hard to find. To improve shelf life and stop explosions, the microbes need to be removed after fermentation and companies use ultra-filtration and short pasteurisation techniques, and sometimes both.

When buying water kefir or tibicos, avoid the usual suspects with artificial sweeteners or too much sugar (over 1 teaspoon or 4g per 100ml) and identify brands that use real tibicos grains rather than just commercial cultures. In the UK, Agua De Madre is a small brand I like. Try and choose live over pasteurised products. Drinks in cans or in bottles with long shelf lives are nearly always dead. But remember, dead tibicos may still be good for you. A sediment in a bottle may indicate it hasn't been finely filtered but, unlike kombucha, is no good to use as a starter.

### *Tibicos in five*

1. They contain between thirty and forty microbes, mostly probiotics and yeast (if alive).
2. Animal and lab studies suggest tibicos could still be beneficial even if pasteurised.
3. Tibicos grains grow fast in sugar but are delicate and easily damaged or killed.
4. They make refreshing alternatives to dairy kefir if you are vegan or lactose-intolerant.
5. They are easy to make and very versatile in terms of flavours and fruits that can be added, being milder and less acidic than kombucha.

## Krauts and kimchis

### How to make simple sauerkraut

- Take one white or red cabbage, remove and set aside the outer leaves and then finely shred the cabbage.
- Add to a bowl with 1 tablespoon of caraway or fennel seeds.
- Weigh the shredded cabbage and seeds and calculate 2 per cent of this measurement to get the salt quantity. Massage the salt into the cabbage.
- Pack down hard into a 1–2-litre Kilner jar, leaving 5cm headroom.
- Use the outer leaves and/or weights in the top to submerge the cabbage below the brine liquid.
- Wait for 7–10 days, burping each day. When it tastes good, transfer to the fridge, ready to enjoy.

Traditionally sauerkraut is fermented slowly under lowish temperatures: 15–20°C for 2–4 weeks. This gives it time for the sourness to develop, and at a low pH of 3.5 it is around seven times more sour than the equivalent kimchi. Red cabbage takes longer than white to ferment, and both work faster if you add some other vegetable (like a small carrot or two).

I think I first tasted sauerkraut in Munich aged eighteen at a street market en route to the Tyrolean Alps for my first prestigious job as an *abwäscher* in a ski resort hotel. This translated as kitchen porter or washing-up boy and was the lowest of all the jobs available – unsurprising as I had no experience of kitchens, my German was terrible, and I was a foreigner. But it paid well and as well as falling for one of the locals chambermaids, I got to eat unlimited amounts of excellent but filling Austrian mountain food. In fact I liked it so much I gained over a stone (6kg) in four months. The staff meals were very generous, and most days we had the option of sauerkraut, usually with wurst, salamis or other meats, and always plenty of mustard and pickles. I got to love the sour, tangy flavours that complemented the meat and the rye breads. Now, whenever I travel in Germany, Austria or find myself in an Alsatian restaurant, I will pick the sauerkraut dish to remind me of those moments. Although I do remember helping transport large vats of sauerkraut, I don't remember us ever making it from scratch and never thought one day I would also be extolling its health benefits. But it is very easy to make and – needless to say – very good for your microbiome.

Fermented vegetables really are the simplest of foods. Sauerkraut is made with two ingredients, cabbage and salt, with time and naturally occurring microbes doing the rest. Many other vegetables can be fermented (or added) the same way. Sauerkraut (which means sour cabbage in German) is the basic ingredient, but the French use the word choucroute, both as a direct translation and a special Alsatian dish where white wine, juniper berries and pieces of meat are mixed and served together, known as *choucroute garnie*. The common story is that sauerkraut came from China and the East around the thirteenth century, when Genghis Khan invaded Eastern Europe, bringing it with him. In China fermented cabbages had been used as food and medicine for centuries – I remember on a recent trip to Shanghai visiting a food market and seeing buckets of fermenting cabbages for sale, all of different maturity and textures. The reality is that fermented cabbage in some form almost certainly existed in Northern Europe before Genghis but never made it into popular culture.

Although they are now associated with German cuisine, fermented cabbages are found in many other countries that may have used them first. We see historical evidence of them further south in Romania, further north in the Baltics, Poland and Russia, and further west in Alsace. German immigrants brought them to America, where in New York they were integrated into the classic Reuben sandwich, and an enterprising Henry Heinz began his empire by selling sauerkraut and horseradish door to door in Pennsylvania. But the kraut love-fest ended due to the First World War, when the ferment became synonymous with a nickname for the German military; although its name was briefly changed to 'liberty cabbage', its popularity faded for a hundred years and eating it was seen as 'non-American'.

In the late eighteenth century it became famous as a medicinal cure and prevention of scurvy. Many years before, in 1747, a Scottish navy doctor had shown that citrus worked in a small, controlled trial at sea – and one of the first epidemiology studies. A group of twelve sailors with scurvy was divided into six pairs. Some were given oranges and limes, or cider and vinegar, which didn't sound too bad, but some unlucky 'volunteers' were randomised to drinking sulfuric acid and seawater in a bid to cure the putrefaction of the guts that was believed to be the cause. Only the two sailors on citrus recovered, although the cider group were happier than the others. Dr James Lind never fully publicised his trial results, which were buried in a large book. The medical establishment at the time refused to believe the new evidence and continued to insist that scurvy was due to putrefaction of the guts, prolonging the deaths of tens of thousands of sailors. Some might say the same attitude to change in nutrition is true today as traditionalists defend the healthiness of artificial low-fat ultra-processed foods.

Twenty years later it was Captain James Cook, who, knowing it was vital to stop sickness and death on long voyages, took a number of anti-scurvy medicines on board including some boiled up limes based on Lind's work. Luckily, he also took barrels of sauerkraut, as he found it lasted better and longer than limes and saved countless lives. It has a high vitamin C content, which in normal cabbage is dramatically reduced by heating (as also happens with limes). His crew were reluctant to eat the unfamiliar fermented dish, and so he introduced a ration system, giving it only to the officers for the first few weeks. It soon became a must-have item as the rest of the crew got FOMO and wanted their ration. The sauerkraut provided many other nutrients, especially compared to the hardtack biscuits riddled with maggots or the rock-hard salted beef, and saved many lives on Cook's voyage.

Both sauerkraut and kimchi are wild ferments, meaning they don't need any starters to kick them off, other than salt. The microbes involved are virtually the same at the beginning, as they live on the cabbage leaves, waiting for their big break in life, and swing into action when they are deprived of oxygen, giving them an advantage over their rivals. In the case of kimchi, scientists recently found that more microbes actually reside on the garlic than on the cabbage, so this was a really useful addition to get fermentation started, particularly in winter. It is also a useful tip for anyone doing wild fermenting: just add a bit of raw garlic as a starter to the mix. This is why wild garlic ferments so rapidly when you add salt.

Technically sauerkraut and kimchi are called synbiotic as they are a mixture of *prebiotic* high fibre cabbage plus live *probiotic* microbes that can live off the cabbage and extend their lifespan. This means that they have microbial advantages, passing through the acidic gut lining where the microbes can keep feeding off the fragments of cabbage and use the minerals and nutrients for energy. The main players in both wild ferments are anaerobic LAB microbes, especially *Leuconostoc, Weissella* and *L. plantarum* at the beginning, to produce lactic acid and carbon dioxide; then as the pH drops and most of the hard work is done *L. brevis* and *Pediococcus* take over. The quantities of microbes you find in these ferments can range from 10 million to a billion bacteria per gram. Some kimchis in particular can deliver over 30 billion microbe colonies in a serving.

## How to make simple kimchi

- Slice any type of cabbage or daikon (Japanese radish) into thin 5mm slices or thin half-moons for radish.
- Add to a bowl with 2 per cent salt, massage and leave overnight.

- In a food processor blitz garlic, spring onions, Korean chilli flakes (1 tablespoon for every 500g), soy or fish sauce (or fermented miso paste for vegans).
- Mix into the sliced cabbage or daikon, then squash into a jar with leaves at top or weights to keep it submerged.
- Taste in 5–10 days and store in the fridge when ready.

Kimchi is just a more complex form of fermented cabbage and is a very broad description of many different fermented mixtures that have both sourness and hot chilli spiciness. Because of the extra plant and spice ingredients it tends to have a greater concentration and diversity of species than sauerkraut. During fermentation, many factors control which LAB types dominate. External conditions such as fermentation temperature, salinity and the number of plants and ingredients all alter which microbes are seen. Garlic, an essential kimchi ingredient, inhibits the growth of non-essential kimchi microbes because of allicin released from garlic during the initial stages. The LAB microbes in the mix are resistant to its antimicrobial effects, and garlic has at least forty species that it naturally carries under its skin. These speed up any fermentation process. In contrast, the Korean chilli flakes, or red pepper powder, slows the rate of fermentation and is associated with an increased proportion of *Weissella* in kimchi microbial communities.

Studies have reported finding at least sixty species in individual samples, including not only bacteria, viruses and yeast but also another type we rarely discuss: *Archaea*. These look superficially like bacteria but are very different genetically. We still know little about them, except several have been found to be probiotic in humans. With over 200 types of kimchi, scientists have isolated over 4,000 microbe strains, and hundreds of different species. The genes of kimchi microbes have so far been found to be capable of producing the anti-anxiety neurochemical GABA as well as a number of other vitamins, and potentially many more useful compounds.

I first came across kimchi twenty years ago when visiting Seoul for an International Twin Research conference and was surprised to come down for my hotel breakfast and be hit with this weird spicy garlicky smelling concoction that nearly everyone was tucking into. To my shame, I was unimpressed and went to the familiar western-style offerings instead. It took me another ten years of sporadic exposure to really get past the smell and into the flavours. But now I am totally hooked and cannot imagine how I lived without it for so long.

Most kimchi has at least six ingredients including cabbage, salt, chillies, onion, garlic, soy sauce and some fermented fish sauce. It was first written about in thirteenth-century Korea, but unlikely to be the same dish as today because chillies did not arrive from the Americas until the late fifteenth century. Early kimchi may have used ginger, radishes and green onions to give it more of a kick, often in the early days without cabbage. Nowadays Korean chilli has its own identity as *gochugaru*, chunky flakes of chilli without seeds that have about half the potency of Tabasco, meaning it's not designed to be super spicy. In Korea, kimchi is more of a verb than a noun and is used in the same way that we might say to pickle something. Nearly every Korean region has their own speciality, and North Korean recipes are a paler, milder version – vital for survival over winter. Each family has their own household recipe, which varies slightly from any others and the average Korean consumes around 40 grams per day. It accounts for the majority of the fibre intake of Korean adults.

The classic cabbage used to make a classic (baechu) kimchi is called Chinese leaf (or napa cabbage in the US). It is more oval and curvaceous than many cabbages, partly resembling a lettuce. The type of cabbage picked is important for technical reasons as they all have different water contents, meaning the brine produced will vary a lot. Very respectable kimchi can be made with daikon (Japanese radish) as a simpler alternative – or in addition – to cabbage; I used daikon for my first kimchi attempt at home and it tasted great. As the ingredients and strength of the chillies vary, so does the fermentation time, which allows different microbes to have their influence on the taste and texture.

If you can't find Chinese/napa cabbage or daikon, it's fine to use other denser cabbages, though they may produce less brine, requiring you to top up the jar with salty water. Young kimchi can be eaten as a crunchy salad and old kimchi can be fermented for six months or more, often kept in dedicated kimchi fridges. In Korea there are national kimchi days for schoolchildren and villages, kimchi in military rations, the World Institute of Kimchi and a kimchi retail index to mark the cost of living. They have even developed a bacteria-free version for space travel. Importantly for continuing the food culture, every schoolchild knows how to make it. Kimchi pretty much defines Korea.

## Krauts, kimchis and health

There are plenty of positive test tube lab studies suggesting the benefits of sauerkraut, backed up by rodent studies showing they can improve some blood tests and liver detox enzymes, but we have only limited human data. One optimistic observational epidemiology study reported a correlation between sauerkraut consumption and low death rates from Covid-19 in Europe. The key graph showed the highest fermented cabbage eaters in Romania and Latvia had the lowest reported deaths from Covid and the lowest consumers in Belgium, the UK and Spain had the highest rates.

But before you get too excited, there are a huge number of biases that could have caused the result. The most obvious is simple reporting differences between countries could have explained all the differences. There is only one specific sauerkraut study of humans reported so far and that was small. This was back in 2018, where thirty-four Norwegian patients with IBS were treated for six weeks with either 75g of pasteurised or unpasteurised sauerkraut. All subjects reported improvements in symptoms, but their disease scores improved in both groups equally. Both groups changed composition of gut microbes, but in the fermented group, clear increases in the microbes *L. plantarum and L. brevis* were seen that were part of the original kraut mix. So this study suggests that – at least in the short term – sauerkraut works whether dead or alive, although some of the effect could be placebo, because they believed sauerkraut would help them.

After a visiting Korean gastro surgeon worked with us at King's College Hospital on the disorder five years ago, I realised how big a problem IBS is in Korea. As for sauerkraut, there are many positive lab and animal studies of kimchi, but few clinical trials. In 2022, a randomised, double-blind controlled trial gave ninety people three types of kimchi to explore symptoms at high doses. They all consumed a lot of kimchi by western standards (210g a day for twelve weeks) as three large daily portions. Unfortunately, they omitted a non-kimchi control group as they believed volunteers would find this totally unacceptable and refuse to take part. They estimated that the participants usually ate around 50–60g kimchi per day, so they may have had a point. The three types they tested were standard baechu (napa) cabbage kimchi, kimchi with added dead *L. plantarum* in nanoparticles, and what they called 'functional' super kimchi, which is classic kimchi with active foods/chemicals shown in test tube or animal studies to reduce inflammation or help weight loss. These included mustard, pears, mushrooms, mistletoe and some extra live *L. plantarum* as a bonus.

The results of all three kimchis showed improvement in basic IBS symptoms, helping both constipation and diarrhoea. In addition, and importantly being less prone to placebo effect, they saw a reduction in blood markers of inflammation (TNF-alpha) in all groups. They only measured changes in microbes in eight people (they were probably on a tight budget), but these showed increases in fibre-eating Firmicutes (Bacillota) microbes and *Lactobacilli* as expected. Some unhealthy bugs like *E. coli* were reduced but numbers were very small. In summary, it was hard to see any major difference between standard and pimped-up kimchis in this study – they all worked well.

Due to the lack of large clinical trials, and the unwillingness of Koreans to stop their regular kimchi, we need to look elsewhere for evidence. Although indirect, results on probiotic trials in humans with IBS have found beneficial effects of supplements containing LAB similar to those present in these fermented vegetables. As well as IBS, rodent studies have claimed a whole range of benefits on reducing other conditions, such as inflammation, improving immunity or reducing body weight. One Korean clinical study is worth mentioning as it took a common microbe found in kimchi and some fermented meats called *L. sakei* and created a probiotic capsule, which they tested against a placebo in 114 mildly obese Korean men and women with a mean weight of 75kg. They gave them a hefty dose of 5 billion CFU twice a day for twelve weeks and told them to follow a healthy lifestyle. At the end of the period the probiotic group had lost a small amount (0.2kg) of body fat and the placebo group – for unknown reasons – had gained 0.6kg, with similar small effects on waist circumference. They saw no other significant changes and no reports of side effects, but importantly the changes were much smaller than seen in the mouse studies, which is an important reminder that we can't overgeneralise from animal studies.

As Koreans consume far more fermented foods than westerners it was useful to look at a 2022 US pilot study of how Americans fared when asked to eat quite large amounts. They tested eating fermented vegetables (sauerkraut and cucumber kraut) vs pickled (dead) vegetables vs no change in diet for six weeks in thirty-one healthy women. They were divided into three groups and asked to eat half a cup of vegetables per day (equivalent to 100g cabbage or 80g cucumbers) or nothing. Participants were supplied directly at home either with fermented or pickled cabbage and/or cucumbers. The compliance rate was around 80 per cent, with most people managing over 90g per day. The study was small so it was no surprise that it showed no differences in blood inflammation levels, or weight or blood pressure. They did show some changes in the gut microbes, with the kraut group showing greater microbial diversity compared to the controls. The group also had higher levels of two helpful anti-inflammatory microbes *Faecalibacterium prausnitzii* and *Roseburia* that have been shown in other studies to be increased both by *Lactobacillus* probiotics and following a Mediterranean diet. It also provided some insights into the tolerance of Americans for fermented foods.

A few people in each group reported trouble finishing their daily quotas, especially as the study reached the end. Bloating was a common symptom, but the fact that participants knew what group they were in could have affected the results. The greatest levels of bloating (60 per cent) were actually in the pickled veg group; 45 per cent in the fermented group; and quite a high 30 per cent in American controls who ate their normal diets. So bloating directly related to the fermentation – rather than just the quantity of vegetables – appears quite rare, at least according to this pilot study. It is useful to remember in nearly every placebo-controlled study, gastrointestinal symptoms like bloating or reflux are reported by around 30 per cent of the population taking the dummy medicine.

Farting and gas is another complaint when eating cabbage, whether fermented or not. This is often due to the chemical called raffinose, which is not destroyed by fermentation or in the upper part of our gut and causes gas and bloating and sometimes pain. Legumes and beans contain even higher levels of raffinose and symptoms are even commoner. When raffinose reaches the gut microbes in our colon they break down the sugar into gases; some people absorb these easily and others unfortunately don't. While there is some random element to this, I have found that these problems occur mainly when you first start eating these foods – the more I eat beans and cabbage, the more my body gets used to them and adapts. Cabbages actually produce more protective raffinose chemicals in winter, so summer varieties may have less windy side effects.

All fermented vegetables can cause rare problems due to histamine sensitivity (see the chapter on cheese for more information on histamine reactions), or get contaminated with unwanted microbes and cause food poisoning. Some Korean outbreaks were traced back to water contaminated with *E. coli* used to add to the salty brine in large or communal kimchi making. US public health experts warn against home fermenting but forget that food poisoning from US restaurants has a much higher rate of infection (especially with *Salmonella*) and that the kimchi scares are a result of the massive consumption of this largely homemade product in Korea.

Another common complaint is that kimchis and krauts contain too much salt, and many western nutrition books advise us to rinse them before eating. One study looked at eating trends over nineteen years in 76,000 Koreans and found that as they swapped fermented vegetables and kimchi for pickled varieties, and kimchi intakes dropped by a third, the total salt intakes also fell significantly from 5.5 to 4.7g a day. As medical dogma states the two are well correlated, this should mean that rates of hypertension (high blood pressure) had fallen, but over the same time period the prevalence of male hypertension increased from 32 to 35 per cent, whereas in the US where hardly anyone eats fermented foods, over the same timescale it also increased slightly from 27 to 30 per cent. For Koreans, despite ingesting an average of a cup (114–164g per day from 1998 to 2016), and the main source of salt in the diet being kimchi and fermented foods, this does not appear to be the main cause of hypertension. Support for this comes from the Korea National Health and Nutrition Examination Survey (KNHANES) where they found no association between blood pressure and high rates of kimchi eating in 20,114 men and women.

One reason for not finding a link with salty kimchi could be the high potassium levels found in cabbage kimchi having a more powerful beneficial effect on blood vessels than the negative effect of the sodium in salt. This mirrors recent studies showing that increasing potassium intake via plants or artificial salt with added potassium has roughly three times the effect on reducing blood pressure compared to salt reduction alone (and is a lot more pleasant). Kimchi and sauerkraut both contain around 160mg potassium per 100g portion, which is only 5 per cent of average daily intakes, or 15 per cent if you eat Korean portions! Eating the salt, potassium and microbes at the same time could have some greater multiplicative effects. Probiotics carried on the cabbages (*L. plantarum*) have been shown in mice and in seven human studies to also reduce blood pressure by a small but significant amount (1.7mmHg). In any case the benefits of eating salt-fermented vegetables appear to outweigh any risks or side effects, but if you are particularly sensitive to salt, feel free to rinse them briefly in water. Who knows – in the future we may be able to buy potassium-rich salt that is also good for fermenting.

Let us not forget that the underlying vegetables, regardless of being cooked, pickled or fermented, are also healthy and provide plenty of vitamins that stave off scurvy. Another bonus are several B complex vitamins, due to the fermentation process improving access to the nutrient. A small cup serving of sauerkraut (100–150g) provides 1.6g of protein, 2.5g of fibre and 2g of sugar, with very little in the way of calories. So despite the lack of any large studies yet, I think the fact that krauts and kimchis are packed with fibre, bioactive compounds and probiotic microbes means that they are likely very healthy for us and our gut microbes.

Whether we need to eat as much as the Koreans is still unclear, but they are still one of the healthiest and leanest developed nations with only 5.9 per cent obesity in the adult population compared to 40 per cent in the US and 26 per cent in England, so it may be a natural counterbalance against the rise of junk foods in wealthy countries. Unfortunately sauerkraut doesn't appear to have benefitted the Germans as much – their obesity rates are close to those of the UK. Their excuse is they don't eat nearly as much cabbage as Koreans (a miserly 1.2kg compared to 25kg per person annually) and German favourite brands (like Hengstenberg) are mostly pickled in wine or vinegar, unlike in Korea.

## What to look for when buying sauerkrauts and kimchis

Usually the healthiest versions are the ones you will make yourself, but you can find good artisan varieties if you look around. It is important to check first that they are not just cooked and then pickled in wine, vinegar or another cheap acid, which would lack any microbial benefit. They can also be fermented with microbes initially but pasteurised before sale to give a longer shelf life. Under crazy labelling laws in the EU and US, manufacturers are not obliged to tell you if the product is dead or alive. Producers of live products will try to give clues like 'unpasteurised', 'raw', 'contains live cultures' or 'probiotic', but these claims are never checked in most countries. The real deal product with live microbes should always be found in the fridge in shops to slow down fermentation and will have a shelf life of a maximum of six months before it goes sour and mushy. When you open it, if it fizzes or overflows it's a sure sign it is alive, but if it doesn't, it doesn't mean it's not. Once open you need to eat it within two weeks, even if kept in the fridge, for the best taste.

Most of the popular sauerkrauts are fermented but then pasteurised, but many popular ones are mixed with white wine, which has the same effect as vinegar. Unlike kimchi, sauerkraut is often eaten heated up; Germany's most popular brand Hengstenberg advises you to cook it for 15 minutes in a saucepan, which would kill off any bugs, as well as reducing many nutrients and vitamin C.

It is good to see that you can now buy locally sourced real fermented kimchi in many places outside East Asia. My current favourite is King Kong Kimchi, made three miles from my house by James Read who wrote an excellent book on kimchi. Another, with a wider distribution, is a Hungarian family business called Vadasz, which makes sauerkraut and kimchi. In our survey the commercial chilled kimchis and many mixed krauts consistently came out top in terms of diversity, with between twenty and eighty-seven species. Beware of cheap options. Price is a big indicator of the quality; there is a sevenfold difference between traditional Korean kimchis and Chinese imports that are growing in popularity. In Spain, desperate for kimchi, I bought a Chinese kimchi that was so spicy it nearly blew my mouth off. There was no translation for my poor mandarin – but with a three-year expiry date, I now know it was pasteurised and I probably missed the warning extra spicy sign.

## Timchi and other fermented vegetables

Both sauerkraut and kimchi are made by dry salting, where you add salt directly to the plant, massage it in and let the water slowly escape. Plants with tougher structures don't produce enough water for the brine and are best cold cooked by adding brine to them. Baby onions, carrots, cauliflower, red peppers, green tomatoes and virtually any other firm vegetables can be fermented by pickling in brine, simply by adding 2 per cent salt to the total weight. The plants that work best tend to have a structural firmness that means they don't rapidly turn to mush. I learned the hard way that veg like aubergines or courgettes are very hard to ferment. Once you realise it is just about adding the right amount of salt and keeping the vegetables submerged in brine and away from oxygen, it's really very simple – the microbes do all the work. If I know I won't be able to eat all my vegetables in the next few days and they do not freeze well, I try not to let them go to waste by turning them into kimchi.

While messing around on social media talking about food waste and ways of reducing it, I decided to put words into action and create my own version – 'Timchi'. Just before leaving for a week's holiday, I took the contents of my bottom drawer in the fridge, cut off any dodgy bits, chopped them up and put them in a mixing bowl with chillies and 2 per cent salt, demonstrating that I could make it in under 10 minutes. I did manage this but be advised: when you rush it can get messy. I squashed it into my Kilner jar with a lid and valve to self-burp itself, came back eight days later and found (apart from the soggy aubergine pieces) that it was perfect. I've included more simple kimchi and other fermented vegetable recipes to try at the end of the book.

## Other ways to enjoy krauts and kimchi

- Added to salads
- Stirred into rice or grain dishes before serving
- Added to cream cheese as a spread – great for kids
- As a Reuben sandwich with pastrami or mushroom pate on rye bread
- With a plate of cheese or toasted kimchi sandwich
- As pie or as dumpling fillings
- With macaroni cheese
- With scrambled eggs

### *Krauts and kimchis in five*

1. Sauerkraut is incredibly easy to make yourself with just cabbage and salt.
2. Both krauts and kimchis have likely health benefits if eaten regularly.
3. They both contain multiple species of probiotic LAB bacteria and yeasts.
4. Shop carefully as shop-bought krauts and kimchis with long shelf lives or stored at room temperature will not contain any live microbes.
5. Fermenting leftover veg is a great and healthy way to reduce food waste.

# Cheese

## The easiest cheese you can make yourself

Labneh-style cream cheese is simple to make and is great if you have made too much kefir or yogurt.

- Take a large pot of kefir or natural Greek-style yogurt.
- Pour it into a sieve lined with a coffee filter, a clean muslin cloth or kitchen paper and set over a bowl.
- Let it drip overnight in the fridge or a cool room to collect the clear whey fluid. (Note: if still very wet, strain twice.)
- Mix your solid curds in a bowl with seasoning, herbs and spices and keep in fridge.
- Keep the probiotic whey in the fridge as a high-protein starter.

While Turkish shepherds were making kefir, others wanted something more solid to eat and transport, something that would last for weeks, not days. They worked out that by allowing curdled milk to dry out then leaving it to cool in a cave, tasty, long-lasting cheese could be made. The lower the water content the longer it would last without getting attacked by stray moulds. Cheese contains protein, fats, minerals and calories, as well as live microbes, and can sustain a travelling human for weeks. This simple preserved food has evolved into something eaten all over the world, now a part of our daily diet and available in a wide range of forms, tastes and costs. Today, if you fancied some raw milk Cabrales cheese aged for several months in a damp cave in the mountains of Asturias, you might have to pay 30,000 euros for a single 2kg chunk.

I once visited a small artisan cheese-making farm in Costa Rica with cheaper offerings. They kept goats and made their own variation of feta cheese, a traditional Greek cheese usually made with sheep's milk (with some goat) and bathed in a salty brine. The owner explained they used a standard yogurt starter of *L. bulgaricus* and *S. thermophilus* and produced lovely cheese. They did not know exactly which microbes grew after the starter to add to the flavour.

A Greek study compared the microbe content in industrial feta with homemade feta from raw milk. While the industrial fetas contained *L. bulgaricus* and *S. thermophilus*, the homemade ones were dominated by *L. lactis* and had several other unwanted guests, including *Listeria* and *Enterobacter*, which at high levels could be pathogens but were luckily at very low levels, being suppressed by the other microbes. An Iranian study of homemade raw milk Lighvan cheeses found more worrying levels of *E. coli* that were present in the ripening stages of the cheese that then dropped off rapidly as it became more mature. This is a problem in homemade Mexican cheeses called Queso Fresco that are a major cause of food poisoning in the US. This shows the importance of looking after cheeses properly and not eating them too early.

In terms of taste differences I have always preferred raw milk cheeses, but is this because I am biased as they are more expensive? The researchers in the Iranian study found that despite the low levels of *E. coli*, volunteers preferred the raw milk to industrial versions. Tasting is often subjective and could be influenced by many other factors so one group of researchers used some electronic gadgetry to simulate whether there is really a taste difference. In the lab they used an artificial electronic nose and tongue and saw clear differences in taste profiles of cheeses made with raw milk and Cheddar cheese made with pasteurised milk in near identical conditions. This proves that robots can tell the difference in flavour and means we should probably be able to as well.

Raw milk contains many microbes that rarely cause harm if drunk fresh but can in some cases proliferate; or cause harm if left too long – tuberculosis is a good example as is *E. coli*. Raw milk is banned in many countries and states; for example it is legal in England and Wales but illegal in Scotland. Most countries allow farmers to drink their own raw milk. As the risk from cheeses being infected with pathogens reduces with time as the cheese dries out, countries like the US and Canada have a strict rule on only permitting raw milk cheese if it is more than sixty days old. Australia had a virtual ban on all raw milk cheeses until, after lobbying in 2022, it finally allowed a range of British raw milk cheeses (mainly aged Cheddars or other hard cheeses) to be imported. So many people have never had the chance to enjoy a lovely, complex raw milk soft cheese like Brie, only having tried mass-produced versions.

The UK, after decades of decline, is having a cheese renaissance and has an estimated 800 different types, possibly more than the French, although naturally this is disputed. Traditionally matured in caves, cellars, stables and mills, there is no one right way of making cheese, which is why it has an infinite amount of flavour possibilities. Cheese like Cheddar is generally made by mixing pasteurised milk with a starter culture of a few species of lactose-loving bacteria (*L. lactis* and *S. thermophilus*) plus an enzyme that cuts up the protein curds. This is either traditionally found in the stomach of young calves or is now made artificially by microorganisms. The liquid whey is drained off and the semi-solid curds are cut into strips, salt is added, and the mix put into round moulds for pressing. They are then left out in a cool cellar (or cave) for a few months, with the surface rinds allowed to grow a mould to help the maturing process and improve flavour. Depending on the levels of acidity and moisture reached at the point of moulding and the speed at which this happens you will have a very different cheese type. British cheeses like Cheddar or Stilton are typically acidic and low in moisture, soft French Camembert is low in acidity and high in moisture, alpine cheeses like Comté are dry but have low acidity, and smelly runny Époisses is both acidic and moist.

While some raw milk cheeses use starters others are avoiding starter cultures altogether, by mixing the new milk with the old batch without washing the pails – a process used in some French alpine cheeses like Salers. These cheeses, because of the sensitivity of natural microbes to tiny changes in the environment, do lack consistency, but have the capacity to be sublime. Cheese microbes keep evolving and changing dynamically over time. A good example is Wensleydale from Yorkshire, which nearly became extinct after the Second World War and was only saved in the 1990s by the intervention of Wallace and his dog Gromit. It was originally a blue mould cheese made by French monks who had moved to northern England from Roquefort in the fourteenth century, then became a white cheese, transforming a hundred years ago into a mild cheese that was easy to spread, and now reincarnated as a crumbly dry cheese.

Within each cheese type (and even within each cheese region) the flavours will vary because of multiple factors, including the plants that the animals eat (grain or grass), the genes of the animal being milked, and the various microbes in the starter culture, milk, and in the farm and caves. The vast majority of cheeses are now made in more microbially controlled ways in factories. Pasteurised milk is collected from multiple high-milk-producing, grain-fed herds and added to commercial rennet and commercial microbe cultures provided by a few global companies. This produces consistency, ideal for high-volume supermarket sales. These commercial starters are used by cheese makers to convert a bland cheese into one of eight cheese styles, such as Cheddar, continental or plain. By later adding ripening cultures you can also alter the final flavour. For example you can select a sweeter, nuttier taste simply by adding the microbe *L. helveticus*, found now in many commercial Cheddars (making them all taste similar). If you want something different, the microbe starter company can suggest other flavours from their catalogue of over 16,000 strains to suit any palate from sweet to savoury, buttery to nutty.

I am not a great believer in drinking raw milk, which is dangerous, but making cheese from it is another matter. A review in 2022 of disease outbreaks due to dairy products in Canada and the US from 2007 to 2021 found only twenty major outbreaks due to raw milk and fourteen outbreaks due to pasteurised milk, the main culprit in both being listeriosis. There is no equivalent data specifically for cheese, but clearly it is rare and has certainly dropped since the advent of pasteurisation. I would not attempt to make raw milk cheese myself, as you have to be super disciplined and careful with your equipment and processes, usually meaning that raw milk cheese makers are really dedicated. Most of them regularly test their batches for pathogens like *Listeria* and accept that each batch will never quite be the same, lacking the consistency of the more common pasteurised cheeses where the starter cultures are tightly controlled.

## Cheese and health

A cheese such as Cheddar is a fatty product: it is 33 per cent fat, of which 23 per cent is saturated, 9 per cent monounsaturated and 1 per cent polyunsaturated. It is also a high source of protein (25 per cent) and has calcium and other nutrients. This amounts to a high calorie density so, by traditional thinking, that should make it unhealthy. But we now know that not all saturated foods are bad for us and cheese that is not highly processed contains live microbes that could be beneficial. But what does the data say?

Most studies have not separated out cheese intakes in terms of good-quality versus poor-quality cheese that may be full of preservatives and lack viable microbes. A recent review of half a dozen epidemiological follow-up studies with combined cheese data (good and bad) found no negative effects on mortality or heart disease from cheese eating, with around half showing slight reductions in heart disease. So in modest amounts we can conclude that cheese is healthy to eat for most people. But some people, as we know from the ZOE PREDICT studies, are susceptible to fats and have to be more careful.

What about other health problems? We hear a lot about histamine intolerance or HIT, which I was sceptical about but turns out to be a real phenomenon. It is a non-allergic reaction in some people to excess histamine levels in food that they can't break down adequately. Histamine is a natural chemical the body uses in times of stress to cause inflammation to help wound healing. Fish and seafood are the most important causes of HIT (especially mackerel and tuna) and have clear safety levels of maximum allowed histamine. Cheese is the next commonest food group associated with symptoms in sensitive individuals. The symptoms come on a few minutes after eating and include sweating, headaches, nausea and vomiting, diarrhoea, itching and burning, swelling and skin rashes. It can also cause increased heart rate, palpitations, fainting and asthma-like attacks, and affects as many as one in a hundred people, who are most commonly middle-aged.

Most people are not affected and can deal with quite high histamine levels by deactivating it with enzymes. But a gene variant has been identified that can increase or decrease your risk of not being able to deactivate the histamine with the enzyme diamine oxidase, and even if you don't have the gene and low levels of the enzyme, 20 per cent of Europeans take medicines that can increase histamine levels, such as some antibiotics, blood pressure medication and some antidepressants. But only recently has science isolated the probable microbes and environmental conditions that produce the histamine. It can occur in both raw milk and pasteurised cheese, but histamine levels are higher on average in raw milk cheeses, and higher in harder, drier, more mature and acidic cheeses like Parmesan, Pecorino, Gruyère, Cabrales; it's also higher in sheep compared to cow's milk cheeses. But there are exceptions and some Brie and feta varieties can also have high levels. These affected varieties can have up to 400 times more histamine than other cheeses. The main microbe causing the problem appears to be *L. parabuchneri,* which sneaks in after the starter cultures, helped by some industrial practices or occasionally sloppiness. Some yeasts have also been implicated, but it has not been easy to eliminate.

Traditional industrial-style Cheddars and similar cheeses will only contain the two or three lactic acid microbes added to the starter culture, plus a few to give flavour at the ripening stage. You can produce extra microbes in cheese in several ways, either by using raw milk, backslopping from previous batches, using an artisan approach to equipment, washing the rind, and adding fungi and yeasts.

## What to look for – and avoid – when buying cheese

In 1953, Canadian James Kraft changed cheese for ever. He had invented ways to superheat and sterilise cheese, meaning it had a longer shelf life and could be moulded without the fat separating. The resulting Kraft dairy slice (or singles), launched in 1965, became one of the bestselling convenience foods in history – gone was the wonderful variability of traditionally made cheeses. As an experiment I have had a Kraft slice sitting on my kitchen shelf for the last five years. Although a bit dehydrated and curled at the edges, it is still spotless and shiny. It may well outlast me. On the other hand, Philadelphia cream cheese, a product line purchased by Kraft, looks highly processed but actually contained microbes. The main bacteria we found were two species of *lactococcus* plus several others in small amounts. Reassuringly, it developed mould after a week or two. A good live cheese should ideally have a characteristic smell and a softness depending on the water content; if left at room temperature it should slowly grow and expand across the plate. A good Brie left out can seem to grow within an hour or two and continue to evolve in smells and flavours. Real cheese eventually produces unpleasant ammoniac odours and will also grow moulds, as the resistance to other bugs reduces over time.

Following a kickback against dietary fat in the 1970s, other cheeses made from skimmed milk entered the scene. These low-quality, tasteless products were mostly added to nutrient-free ready meals, which, once cooked, were devoid of microbial life. Using a vast suite of chemicals, flavourists can now bypass microbes and recreate the essence of Cheddar, Camembert or blue cheese. These flavours are used to liven up cheap, bland cheese or replace cheese entirely in savoury snacks. These extra flavours are also added to low-fat cheeses and other ingredients to replace the natural flavour and texture of cheese fats. Despite their best attempts, they are often bitter and rubbery.

### *Pizza cheese*

This deserves a special mention. Mozzarella is the most popular pizza topping, traditionally made artisanally from herds of local buffalo. Although buffalo mozzarella is now made worldwide, the original and best is Mozzarella di Bufala Campana, which is made in a (DOP) protected area of southern Italy. But where there is money, there is organised crime: Giuseppe Mandara – the self-declared 'Armani of Mozzarella' – used the business to launder money until he was arrested for passing off cheap provolone cheese and contaminated products. In 2016, genetic tests still showed that one in four 'local' mozzarellas contained foreign cow's milk.

In a world where one in seven Americans is eating a pizza slice as you read this, the mozzarella used for most pizzas today is rather different. It is often sterilised, frozen and mixed with cheaper cheeses. Frozen pizza is designed to last for years in a $150 billion world market underpinned by analogue cheese. Analogue cheese is a term for any pseudo cheese that contains ingredients that mimic or blend with the original at a lower cost. A 2016 survey of British pizza takeaways found one in four were selling analogue cheese as real mozzarella on pizzas and around three-quarters of frozen pizzas contain analogue cheese. Your chance of getting a probiotic hit from pizza is rather slim, so you may want to look elsewhere for gut health.

### *Non-fermented cheese*

A few other non-fermented cheeses are worth discussing here too. Cottage cheese is often tasty but is not really a cheese; it is made from acidifying the curds into chunks and its high protein-to-fat ratio makes it popular with weightlifters. Ricotta, made in Italy since the Bronze Age, is also not technically a cheese, as it is not fermented. It is made from the leftover strained whey of other cheese-making (traditionally pecorino) using acid in the form of vinegar and boiled to high temperature, killing any microbes. This makes it a tasty, low-fat food but a poor probiotic. Its name means 're-cooked' and has popular equivalents in other countries that lack live microbes, such as requesón in Spain, recuit in Catalonia, and paneer in India. Ricotta is a very versatile cheese-like product and commonly smoked, baked, salted or sweetened as a dessert or filling (as in Sicilian cannoli). It is surprisingly easy and safe to make at home by heating whole milk, salt and vinegar (or lemon juice) together and straining the resulting curds. Fermented swaps for cottage cheese are cream cheese, and Greek-style yogurt works well in place of ricotta, although if adventurous you can add microbes to them yourself in the form of brines or probiotics.

### *Vegan cheese*

Mass-produced vegan cheeses are popular, relatively bland and easy to eat, made from coconut oil, starches, added protein with lactic acid flavourings but, sadly, no microbes. They are the starchy equivalent of ultra processed analogue cheese – but without the animal and environmental costs. In the UK, blind taste tests reveal that consumers often prefer vegan cheese slices to the real equivalent, at the lower end of the ultra-processed market.

Non-ultra-processed vegan cheeses are slowly improving, however, and I prefer the ones made using nuts and nut oils. One form of probiotic vegan cheese that has become relatively popular is fermented cashew nut cheese. You simply soak cashews overnight, add salt and mix to a paste in a blender, then add a probiotic starter and ferment for twenty-four hours. With just a handful of ingredients, you can create something that roughly approximates the texture of real cheese.

### *Cheese in five*

1. Traditional cheese contains live microbes and is probably slightly healthy for most people, despite the high saturated fat content.
2. Fake or ultra-processed cheese is unhealthy, contains no microbes and should be avoided.

3. Raw milk (unpasteurised) cheese contains a greater diversity of microbes, often providing richer variable flavours.
4. Vegan cheeses are often made cheaply from coconut oil or palm oil with added chemicals and no microbes, but some nut-based ones are much better.
5. You can make your own cheese easily from yogurt, kefir or by blending nuts.

# Kombucha

## How to make simple kombucha

- Make a large brew of good-quality black or green tea (8 tablespoons of leaves or 4 teabags to about 75ml of boiling water). Let it cool and dilute it with cold water in a 2-litre glass container.
- Add good-quality white sugar at 7 per cent of the total weight of water (140g for 2 litres of water).
- Once cooled to below 35°C, add your SCOBY.
- Add a splash of an old mature kombucha brine or bottle and cover with a loose lid.
- Wait for roughly 9 days, or until you have the sourness you want.
- Transfer the SCOBY back to the fridge hotel with some kombucha juice to keep it alive.
- Strain and bottle your kombucha and chill in fridge to enjoy.

I tasted my first kombucha around 2010 while attending a medical conference near Bristol. A local company had been making what they called kombucha beer with 1 per cent alcohol and were offering it free to delegates. Most doctors enjoy a freebie, and I am no exception so I was keen to try it. I'd like to be able to say I was hooked from that day, but actually I thought it rather sour and unpleasant. I stayed clear of it for several years, until I started tasting fruitier versions and then tried making my own. Although it is a taste that takes time for most people to get used to, it is definitely worth it.

Kombucha is a versatile fermented tea, generally made with black (or sometimes green) tea, with many added flavour options. It usually has less than 0.5 per cent alcohol and is regarded as non-alcoholic, with low levels of caffeine. The starter is a kombucha SCOBY (symbiotic culture of bacteria and yeast) or mother, a rubbery slippery disc that you add to a flask of room temperature sugary black tea and wait 1–2 weeks for the microbes to do their job and produce some sourness, fizz, small amounts of alcohol and a radical change to the flavour. Many extra volatile flavour compounds are produced, with studies showing two to three times more volatile flavour compounds (up to fifty-six in some cases) after ten days of fermentation. The result is something that is totally distinctive in taste and flavour, which I now love. The complex flavours come from the equally complex group of microbes and their genes and enzymes that have evolved to live together as a collective colony and feed off the tea and sugar. An optional second fermentation phase is often used to add more fruits and spices and a fizzier, more palatable product.

The SCOBY comprises a mixture of many acid-forming bacteria, the *Lactobacillus* LAB group, the acetic acid AAB group and a near infinite number of different yeasts, which vary considerably between batches. A Turkish study used gene sequencing of several samples and claimed the dominant bacterial genus as *Komagataeibacter* and dominant fungal genus as *Zygosaccharomyces* in all samples. Other studies have looked at commercial brands and found rich microbe diversity, but no consistency in the dominant strains, which also included *Bacillus coagulans, L. nagelii, Gluconacetobacter, Gluconobacter* and *Komagataeibacter*. Our own survey could not extract microbe DNA from several larger brands in cans, but found between eighteen and twenty-four common bacterial species in the smaller bottles from UK brand Momo. The microbes break down the tea and sugars to produce more acids, carbon dioxide and a small amount of alcohol as well as a wide range of bioactive compounds such as vitamins. In addition, the acetic acid bugs create more slimy cellulose so the blob can keep growing as all the microbes reproduce. We are just starting to realise the staggering range of chemical compounds produced. As well as sugars we see over a dozen organic and phenolic acids, water-soluble vitamins B1, B2, B6, B12 and vitamin C, eighteen essential minerals like iron and magnesium, over twenty polyphenols, lactones, amino acids, as well as purine alkaloids found in the tea like caffeine and theophylline.

Kombucha is also known as 'Manchurian' or 'Kargasok' and, as you can guess, probably originated in north-eastern China or Manchuria around 2,000 years ago. The actual origin of the mysterious SCOBY is unclear, although there is a plausible theory that fruit flies fed off some abandoned sweet tea somewhere in China and added the contents of their guts that contained acetic acid and lactic acid bacteria with some yeasts that were floating around. Another theory is that it was popularised by a Korean Physician (Dr Kombo) who travelled to Japan to treat Emperor Ingyo's gut problems in 414 BCE. From Japan and China it made its way to Russia where it became popular, and finally in the twentieth century the drink filtered into Europe. In the US it gained popularity, especially in California as the HIV pandemic hit, as it was believed to have immune-boosting powers. It is now a global $2 billion market and growing, with companies like PepsiCo buying out smaller companies. There is clearly a big future market for this microbial soft drink, which – like unfermented tea – is very adaptable. As well as different fruit, herbal and even coffee versions, there are now water kombuchas, alcoholic cider and beer options, known in the USA as hard kombuchas.

The simple process using a starter mother is similar to vinegar-making, as we'll see later, but it is a slow process, unlike tibicos. The SCOBY is a jelly-like community of microbes that form a strong biofilm structure made of cellulose, binding them together to protect each other. This is also known (incorrectly) as a tea fungus or by its biological name *Medusomyces* that means jellyfish fungus. For most people it is one of the more mysterious ferments and has an aura of danger about it that puts people off. But I was intrigued and it was one of the first ferments I got into.

I first bought a small commercial bottle in my local pickling store. I chose one that had a small circular sediment in the bottom. I drank most of the bottle, found I enjoyed it and kept the baby blob alive with some sugar added to grow it into a bigger baby blob or 'mother' – a SCOBY. I then brewed up some sugary black tea, cooled it in a large glass jar and added my blob. I covered it with a clean cotton dishcloth and left it for ten days. The blob initially sank to the bottom of the bottle as if dead (to my dismay) but then floated to the top and steadily grew to a centimetre thick lid, filling the diameter of the container, physically stopping other microbes gaining access, and preventing them producing acid. Over the next ten days, my blob clearly enjoyed my sugary tea and grew from a baby that was less than a centimetre wide to a strapping adolescent about 15cm in diameter and hard to cut. After three weeks I removed the blob, put it back in the fridge in some kombucha liquid and decanted the kombucha into bottles, adding a piece of ginger to each. I must say I was the only one to enjoy the super sour concoction, apart from a few hardcore and more polite friends, and I have since learned to stop the ferment earlier before it becomes vinegary. I know now to use a pH meter, as well as my taste buds. My first batches had really low pHs of 2.2–2.8, but my more recent ones are 3.5–4.0 and taste much more pleasant.

One of the great things about kombucha is that you can see the fermentation process in real time, with the bubbles, acid and new aromas changing daily. The microbes are in charge and the only thing you can do is decide (by trial, error and tasting) when to halt the fermentation, either by cooling it down or removing the blob. Because you can't always control what microbes are growing in your blob, some unwanted strains can occasionally take over, and I occasionally have to perform some mild scraping of my blob to help the good guys take control again.

I invented several new kombucha cocktails involving Cava and Aperol as part of my research programme, which tasted pretty good, though you need to go easy on the alcohol if you want to avoid killing the microbes, as anything above 15 per cent alcohol kills everything (in theory). But as discussed earlier, we now know that even dead microbes can be helpful. Making kombucha would be an ideal experiment for children to perform in a science class (or at home), as you get to see and taste fermentation in real time and understand the incredible activity going on every day in our own intestines.

## Kombucha and health

The health claims for kombucha are wide, controversial and sometimes way over the top, from treating cancer to curing AIDS. As usual, there are a dearth of human clinical trials for a food product that can't easily be patented and is often thought of as 'complementary medicine'. Kombucha's health benefits come from the tea and the many products of fermentation, including glucuronic acid, acetic acid, polyphenols, phenols, B-complex vitamins and vitamin C as well as the microbes. The list of test tube and animal studies (in mice, rats, chickens, pigs, cattle and rabbits) showing benefits is long. These include fighting infections, liver and gastrointestinal functions, immune stimulation, detoxification, antioxidant and anti-tumour properties, and fighting cancer, cardiovascular disease, diabetes and neurodegenerative diseases.

In 2003, there was a backlash from the established medical community with a highly publicised review of kombucha, essentially saying because there were no clinical trials in humans, it clearly had no health benefits and was associated with serious adverse events and only benefited those making money from it. The review quoted several cases of liver disease, a strange case of anthrax on the skin from someone who used the SCOBY to treat an infected cut, and one fatality. The mysterious way kombucha is made ensured the concerns of the medical profession persisted. In 2017, a case report in the *BMJ* made the headlines. It highlighted the case of a fifty-four-year-old lady who was admitted to hospital with asthma and sickness and found to have very high levels of lactic acid in her blood. The doctors were puzzled until she told them she drank a daily home brew of kombucha. Other cases in the US have reported patients dying of liver failure, although the link with kombucha was never completely proven; another found an immunodeficient patient developing renal failure twenty-four hours after drinking kombucha and another lead poisoning from the lining of the kombucha cask. Some batches can get contaminated with fungi and moulds and in one incident two people who drank it got sick, but 132 others who also drank the same batch remained healthy, showing this is a minuscule problem compared to being exposed to standard food-borne infections or visiting your local kebab shop. By 2019, as drinking kombucha became more mainstream, dentists warned that they were seeing tooth decay and enamel erosion in overzealous drinkers due to the acidity and residual sugars.

But overall the tide of opinion has now turned and a systematic review finally concluded that kombucha drunk in normal amounts (120ml per day) was not harmful. The only caveats were for pregnant women and patients with very weak immune systems, for which (as usual) there was no hard data either way. The review authors still complained there were no proper clinical trials to confirm any health benefits, apart from one PhD thesis from India that was an uncontrolled study of twenty-four Indian subjects with type 2 diabetes whose blood sugars improved after daily treatment for ninety days.

The paucity of good research changed in 2023, when a well-known Sydney group published a small but careful randomised, placebo-controlled crossover study in eleven adults comparing the effects of drinking soda water, diet lemonade or kombucha on the blood sugar responses to a high carb meal. They found that kombucha produced around a 15 per cent reduction in blood sugar and insulin levels compared to soda water. This reduction was in spite of the extra half spoon of sugar in the kombucha that the other drinks did not have. So the only human study to date has at least replicated many animal experiments that showed a glucose and insulin lowering effect. The authors couldn't easily explain the mechanism, as it wasn't just the acid effects as seen with vinegars that can slow down the release of starches, because the lemonade was just as acidic. They put it down to the complex interplay of healthy microbes and bioactive factors in the kombucha. The greater the complexity of the microbes and the food they feed off, the greater the range of bioactive chemicals. A recent study found that just adding pineapple peels and cores to kombucha significantly improved its already impressive ability to ward off unwanted microbes. In summary, although the only hard evidence in humans to date shows a benefit to blood sugar control, the profile of the microbes and the large knowledge base from lab and animal studies makes it highly likely this is a beneficial healthy product.

However, if you are not making your own you do have to choose the right one . . .

## What to look for when buying kombucha

There are now a huge range of commercial kombuchas available in most countries, ranging from the increasing number of artisan varieties in the UK, such as Momo (whose fermentary in London I visited, where I saw their genuine blobs growing in vast jars), to industrially made pasteurised kombuchas full of artificial sweeteners and flavourings that are more like a synthetic iced tea that can last years. Artisanal brands can send fresh samples in batches and keep a shelf life of a couple of weeks in a fridge before they turn too sour. As usual when scaling up in size nationally it is hard to maintain quality and long shelf life when transporting a live product across the country, especially in sealed bottles or cans. This is why companies have to keep prices high, cut corners or find clever solutions. Most commercial kombucha companies need to stop fermentation to allow a shelf life of at least six months. They spin the cloudy product and then ultrafilter it to remove particles, which cleans it up but also, importantly, removes microbes. Some also pasteurise it to be sure no live microbes remain. Other companies (e.g. Remedy) rely on long fermentation times of over a month (rather than ten days) so that all the original sugar is used up by the resident microbes. This means the product is now stable and the microbes that are still present cannot continue to replicate and grow or produce more carbon dioxide, making it fizz and explode. They are essentially sleeping or in a coma. The good news is there is virtually no sugar in the final product, which is good for your teeth and reduces sugar spikes; the problem comes from the fact that the drink is very sour, so dilution plus artificial sweeteners like stevia or xylitol are needed and these may have some unwanted effects on your gut microbes. Some companies, such as Equinox, also do long fermentation, and manipulate environmental conditions such as heat in order to selectively keep only the calm microbes that won't produce more acidity or gas in the cans or bottles, even if stored at room temperature. The good news is that most commercial versions – even if they have added sugar or sweeteners – generally still contain a wide range of healthy bacteria and yeast, producing many bioactive compounds that can be healthy, even if in most cases the microbes are more likely dead than alive.

As always, check the labels carefully for the amounts of sugar (carbs) added, which are highly variable as some can be very sweet (up to 5 per cent or 4 teaspoons of sugar per bottle). These will actually give excessive blood sugar rises despite the gut benefits, so the drink may score badly overall on health apps (like ZOE). In general – and there is some debate – artificial sweeteners are healthier than added sugar, though you ideally want to keep both to a minimum. Older artificial sweeteners like sucralose, saccharine, acesulfame potassium and aspartame are especially to be avoided as they disrupt the gut microbes and our metabolism as well as our gut lining and may negate some of the health benefits. Newer sweeteners (often labelled as natural) like stevia, monkfruit or xylitol may be better, although the science is changing fast.

A warning: the information you see on the labels is very patchy, and the claims vary enormously with little control unless someone complains. You are not obliged to say if your product is pasteurised or filtered, or dead or alive. (In the US I saw some of them labelled in small print as 'gently pasteurised', which is like saying the bugs are gently killed!)

The ultimate test of live kombucha is whether your shop-bought kombucha can regenerate itself within a few days when the sediment is put into a flask of warm sugary tea. A small amount of sediment or the signs of a baby blob forming are the best sign of a live kombucha as opposed to one made industrially from a few kombucha cultures that lack the wider SCOBY community. I recently tested some kombucha sold as a concentrated liquid that you dilute and saw no activity for two weeks. I was just about to abandon hope, when a thin and puny SCOBY started to form on the surface, suggesting that at least some of the microbes were present, but not enough to make me want to feed it properly.

There are plenty of kombucha brands out there but remember not to be fooled as they can all use the term artisan, natural or even live cultures as these are rarely challenged. Some may just add the cultures without a SCOBY, so be wary of very long shelf lives and long lists of ingredients and chemicals. A good brand should just have raw kombucha cultures, sugar, tea and some fruits and not taste too sweet.

## Kombucha variants

- Jun is made using honey rather than sugar, which gives it different flavour, but the process is otherwise the same.
- Adding pineapple waste at the beginning is an interesting addition.
- Adding coffee at the second fermentation stage makes an interesting brew.
- Kombucha with fermented hops is a lovely drink, similar to beer in flavour.
- Hard kombucha is with alcohol between 0.5–8 per cent and is popular in the US.
- Leaving kombucha exposed to air for too long will make it undrinkable but turns it into a useful tea vinegar.
- Add 3 tablespoons of kombucha to double cream, leave at room temperature for 24 hours and then in the fridge for another day and you'll have perfect fermented crème fraîche.
- Kombucha can also be used as a starter to make cultured (fermented) butter.
- Fancy a kombucha accessory? Clothes can be made from excess SCOBYs that have grown up to a metre in diameter and 15cm thick, but you may have to fill your bath with tea.

*Kombucha in five*

1. Kombucha is easy to make by adding tea and sugar to a SCOBY mother starter.
2. Contains up to sixty microbe species of LAB, AAB and many yeasts, producing hundreds of bioactive health compounds and vitamins.
3. In humans it reduces blood sugar spikes, though is likely to have wider beneficial effects.
4. Beware of commercial brands with artificial ingredients and preservatives, excess sugar or sweeteners and long shelf lives – and no sediment.
5. Ideally pick artisan brands with sediment and few ingredients, or make it yourself cheaply.

# Vinegar and tepache

## How to make simple wine vinegar

- Decant the remnants of an opened bottle of wine (50ml) into a wide-brimmed glass jar.
- Dilute it with water (20–30ml) so the alcohol content drops to below 10 per cent (ideally around 7 per cent).
- Cover with a cloth to let air in and bugs out.
- Wait 8 weeks to ferment until it smells vinegary.
- Transfer to a narrow-necked bottle with a lid or cork to slow further fermentation.

Not many foods are a prized cooking ingredient, drink, household cleaner and food preservative. The earliest record of vinegar goes back to 5000 BCE in Babylon, when mashed up dates were turned into alcohol and then vinegar. Hippocrates mixed honey and vinegar to help treat coughs and vinegar was undoubtedly pretty widespread anywhere beer or wine were made. The modern word vinegar comes from the French *vin aigre* or sour wine. Roman soldiers drank a diluted version of it called posca, although the best stuff was very expensive. Vinegar can be made from any sugar-containing source that can be made into alcohol, such as white or red wine grapes, rice, potatoes, malted grains and barley (for malt vinegar), various palms (such as nipa from the Philippines) and fruits, such as apples to make apple cider vinegar or pineapples for tepache. It can also be made from kombucha, as I discovered in my first attempts at fermentation when I left it too long and it became like vinegar.

Vinegar can also be made directly from an alcoholic drink, therefore skipping the first alcohol step. But the yeasts can only work if the alcohol content is below 15 per cent, as they haven't yet evolved to survive at the higher level. This is good news for port, gin and whisky, which can't accidentally turn into vinegar and are protected from microbial attack.

AAB microbes in the presence of oxygen ferment the alcohol into acetic acid; this increases the acidity to 4–8 per cent, ensuring that other non-acid-loving bacteria can't survive and compete. In the process, the microbes also produce a range of other complex chemicals and flavours depending on the original plant that was used to generate the alcohol. A vinegar mother is often used to get a new batch started and this is a mix of yeasts and bacteria that form a single blob, similar to but softer than the SCOBY used in kombucha.

Vinegar usually contains at least 4 per cent acetic acid but this may go up to 8 per cent and the strengths are usually referred to in this way rather than by their pH as for other ferments. Vinegar also contains substances that provide potential health benefits and extra flavours and complexity. These flavours range from sour to savoury to sweet. The predominant bioactive compounds with potential healthy properties found in vinegar are melanoidins (a by-product of the Maillard browning reaction that also appears in coffee), multiple polyphenols, over twenty minerals such as calcium and iron, B and C vitamins and alpha-glucans. In Chinese medicine, many vinegars are used but rice vinegars are believed to have the best properties and results, the most common being Zhenjiang aromatic vinegar. The microbes produce the vinegar mother; two complex EPS sugar structures, levan and cellulose, stick together, acting as a sanctuary for microbes. The levan is made of a well-known prebiotic fibre called FOS (fructo oligosaccharide) that has health properties of its own and is a natural emulsifier and bulker often used in the food industry.

As well as the yeast *Saccharomyces cerevisiae* there are several different sub-species of bacteria in vinegar but all are acetic acid-producing, mainly from the *Acetobacter* family and a few *Gluconobacters*; and as usual there are more in the mother than the liquid, but nothing else can grow in this highly acidic environment. Different strains help make the dense structure of the mother. Studies have shown that vinegars have antioxidant and especially antimicrobial activity due to the compounds they produce, which is why in the past they were used to fight infections and heal wounds. You can make vinegars from many plants as long as there is sugar that converts to alcohol, because the *Acetobacter* are floating everywhere in our environment.

## Vinegar and health

As well as an ancient condiment, for millennia and across the world, vinegars were regarded as health tonics, with claims that they reduce blood pressure, liver disease, cancer, heart disease, obesity and blood sugar. The hard – rather than anecdotal – evidence to support this has generally been lacking, although, as usual, there are plenty of laboratory and rodent studies. Looking at these studies it is clear that if you are an obese mouse, drinking vinegar daily will be good for your blood pressure and your blood fat and blood sugar levels. Chicken farmers have fed vinegar to chickens as an alternative to antibiotics to help them grow faster, which seems sensible.

But what about humans? The first study I could find looking at blood sugar in humans goes back to 1995, where scientists took five subjects and fed them a salad with or without vinegar and then observed a 20 per cent drop in their blood sugar levels after they had subsequently eaten bread. A number of other slightly larger studies followed, showing that using 1–5 tablespoons of vinegar had the same effects. The results appeared to be similar in those with mild diabetes and in healthy people, based mainly on tests with carbohydrate-rich meals. The usual studied dose is 20g of vinegar per meal, which is equivalent to 1g of acetic acid or 3–4 tablespoons.

Vinegar, especially rice wine vinegar, is a very common component of Japanese cuisine, appearing in many sauces and in sushi. The Japanese have faced big problems with blood pressure and strokes and so there has been interest in any foods that lower blood pressure, such as rice vinegar. Many laboratory and small clinical studies have been published in Japan from the early 2000s, though these were mostly overlooked by English speakers. There was also interest in brown rice vinegars because the calming neurochemical GABA is found in high levels in brown rather than white rice, which they thought could be useful in reducing blood pressure. GABA levels are further increased by fermentation using LAB bacteria such as some *Weissella* strains.

Studies also showed that GABA-enhanced white rice did have a mild blood-pressure-reducing effect beyond that of normal rice, showing that GABA was important. One study in 2008 of seventy-five people with mild hypertension showed a reduction in blood pressure after twelve weeks of brown rice wine vinegar with high levels of the bioactive compound GABA. But another study added rice wine vinegar to dried bonito and found the effect to be more related to the combination of vinegar and bonito than any added GABA. Although these studies are small, they are at least performed in humans with placebo controls, so whatever the mechanism there is some support for fermented rice vinegars being helpful for blood pressure.

The Japanese have less of a problem with obesity than most countries, but in 2009, 175 obese Japanese men took either 15ml or 30ml of rice vinegar as a trial for twelve weeks against a control group. The study found impressive results, with reductions in body weight, internal fat and blood fat (triglyceride) levels. Strangely, that particular study did not find significant reductions in blood sugar. A 2021 summary combining results of 686 participants across nine human studies taking different doses of apple cider vinegar (ACV) versus placebo concluded that it lowered triglycerides and total cholesterol, but it did not make a difference to LDL or 'bad' cholesterol levels, which are more tightly linked to heart disease. Most studies explored taking 2–6 tablespoons a day. It also showed a consistent effect on lowering fasting blood sugars in those that took it for over eight weeks. The caveat was that some studies were of poor quality and most were on patients with type 2 diabetes; when volunteers without diabetes were considered, the results were of a smaller benefit.

So in multiple studies vinegar has been shown to reduce blood sugar before and after meals by up to 30 per cent and so reduce sugar spikes. Although the effects in people without pre-diabetes might be less impressive, we know these spikes, if persistent and regular, lead to inflammation and increased risk of heart disease, obesity and type 2 diabetes. The studies so far only looked at relatively short-term effects, and we don't know the long-term benefits of vinegar. Not long ago, I would have been shocked if you told me vinegar had any of these effects, so I find the different possible mechanisms of action really interesting, but we don't really know yet if it's primarily the acidity or the fermentation in the vinegar that is key.

One mechanism is that acetic acid in vinegar is detected and slows down the gastric emptying rate (the rate at which the stomach empties food into the small intestine). This in turn slows the breakdown of starches and gives you more time to reduce glucose levels in the blood. Another is that vinegar increases the uptake of glucose from the blood vessels into the muscles. This is where a small amount of glucose is normally stored and vinegar may just enhance this. There is supporting evidence from muscle biopsies in eleven patients with type 2 diabetes having vinegar or placebo, and also in racehorses fed vinegar to improve performance. A third mechanism is that vinegar boosts the production of insulin from cells in the pancreas and this has been backed up by human studies. Finally, vinegar slows the digestion of starchy carbohydrates by interfering with the activity of the enzymes that break them down. These enzymes are mainly in your small intestine, which is after your stomach and duodenum, but they start working much faster in the saliva in your mouth as soon as you start to eat. The main enzyme called amylase is prevented from working efficiently by the acidic vinegar. There are a number of human genetic studies, some I helped with, linking levels of amylase with obesity, so this mechanism is very plausible, especially given the rapid action of vinegar on sugar levels.

As the clinical studies used different forms and doses of vinegar it is hard to know what component was the most important. Was it the acetic acid-forming microbes, the vinegar mother or its products, its antioxidant properties or simply the acidity? The antioxidant activity of vinegar is ten times less compared to most common fruits, so that is an unlikely mechanism. One study neutralised vinegar with sodium bicarbonate to change its pH and found that it then didn't reduce blood sugar, strongly suggesting that the acidity of the vinegar – rather than its constituents or microbes – was the key factor.

The role of acidity and resulting pH is important here. Most cheap vinegars, lemon juice (and even Coca-Cola) have greater acidity and a lower pH (around 2.2–2.4) than expensive vinegars like real balsamic (around 3.2–3.4) or sherry vinegar (3.5–3.7). To put this in context, our gastric juice is much more acidic at around 1–2 pH (remember these are on a log scale), our blood is slightly alkaline at pH 7.4, and toothpaste, which we use to reduce acidity in the mouth, is 9.9 pH. If the pH of vinegar is too low, only *Acetobacter* can survive, but as the pH gets above 3, other beneficial microbes can start to flourish, up to around pH 5, when other unwanted guests can start to survive. So the reason that kombuchas may be better for us health-wise is that as well as having only a mild acidity (around pH 4–5), which helps blood glucose, the acidity is not overpowering for the other beneficial probiotic microbes.

We don't yet have all the data, especially on the gut microbiome, but for the moment it looks like milder tea vinegars like kombucha (if overfermented) have a more generalised benefit on our health than fruit vinegars, though they all still have their place. I haven't seen any convincing data supporting the much-hyped benefits of apple cider vinegar above others, and the only study showing a weight loss benefit used rice vinegar not apples. If you have the money, investing in a real brown rice wine vinegar, or a balsamic or sherry vinegar with a sediment that can form a SCOBY mother, might be worthwhile for your health, particularly if the milder acidity means some helpful bugs survive.

The acid in apple cider vinegar has the potential to cause issues if you don't dilute it properly or use too much of it. The recommended benefits are seen with consuming around 2 tablespoons per day. However, apple cider vinegar could also pose risks for people with certain health conditions that are worth knowing about. These include: 1) Damage to tooth enamel. Over time, the acid in vinegar could erode the enamel on your teeth, leading to cavities. It's best to dilute apple cider vinegar before consuming it and rinse your mouth with water afterwards or drink it through a straw. Crucially, avoid brushing your teeth for at least an hour after drinking vinegar, when your enamel is most fragile. 2) Burns. Undiluted apple cider can cause chemical burns if applied to skin, or, if you regularly drink vinegar without diluting it properly, it can lead to oesophageal ulcers. 3) Vinegar can lower your blood sugar and insulin levels, which could cause significant dips in blood sugar levels if you have type 1 diabetes and are not monitoring them closely. 4) Taking regular vinegar can reduce the amount of potassium in your body, rarely affecting people on strict keto diets or taking some blood pressure drugs.

## What to look for when buying vinegar

Different forms of vinegar have different bioactive compounds that can support health. For instance pomegranate vinegar has more polyphenols and antioxidant compounds than apple cider versions. But it is still unclear how we rank the relative importance of the acetic acid microbes or their products compared to the pH and acidity of the concoction.

Taste is another matter and if you pay extra for fermented balsamic or sherry vinegar you will notice the extra complexity of taste and flavours that you don't find in diluted versions. If you want to guarantee the real balsamic you have to look for the words 'Traditional balsamic vinegar of Modena' and be prepared to pay £40–60 or more for a small 100ml bottle with a DOP sticker, knowing that it has spent anywhere between twelve and ninety-nine years being passed through barrels. On the label '*cooked grape must*' should appear first as the main ingredient. But millions of balsamic vinegar bottles from Modena have magically appeared on sale everywhere at surprisingly reasonable prices. This mass-produced version is made very differently, speeding up the process using mainly red wine vinegar with added wine must and sugar. It is called 'balsamic vinegar of Modena' (with an IGP label) and is found in many supermarkets, costing from £2 to £20 per bottle, depending on how much time it spent in a wooden barrel as opposed to a factory. Having said that, many of these modern copies can taste fine and have some health benefits. But be wary of balsamic vinegar 'glaze' that you might drizzle over your goat's cheese salad: even from Modena, I found some that were 45 per cent sugar and not very healthy at all.

To add to the confusion there are many certifications and labels according to ageing and barrels and region. An IGP label merely certifies that the vinegar comes from the Modena region and has the correct grapes but doesn't tell you much about quality. Outside Europe, the rules are even more lax and virtually anything can be called balsamic from Modena. Even further down the food chain you'll find plain 'balsamic vinegar', which doesn't claim to be from Modena and is a low-quality product made from industrial vinegar (just acetic acid) plus caramelised sugar and variable grapes, and usually a sickly sugary taste. As is often the case, you get what you pay for. A good alternative to balsamic is sherry vinegar, which has some depth and interesting flavours that are worth trying. Other cheaper commercial vinegars you can buy are red and white wine vinegar, cider vinegar of many different qualities, beer vinegar, rice vinegar and white or industrial vinegar (used mainly for cleaning not drinking). Some of these, like cheap balsamic vinegar, are not fermented at all; they use industrially produced acetic acid plus water.

Making your own vinegars is very simple and is a great way to avoid waste. Leftover bottles of wine from a party just need to be diluted to below 10 per cent alcohol to allow the acetic acid-forming microbes to perform, then be exposed to air. The skin and waste products of apples are a great example, which you'll need to mix with sugar to create alcohol and then wait for the alcohol to turn to acetic acid and apple scraps vinegar, all thanks to the microbes. Another fun way is to use the leftover peel and core from a fresh pineapple to make tepache, a traditional Mexican drink that has been around since pre-Columbian times that is slightly alcoholic and both sweet and tart. It is sold by street vendors and found commercially in Mexico and California in cans and bottles and drunk as a non-alcoholic soda, but beware the high sugar content.

## Making your own tepache

To make your own tepache, simply mix leftover pineapple scraps, including the skin, with 5 per cent (about 150g) unrefined cane or brown sugar, then leave in a jar for a day or two for the natural yeast (*Saccharomyces*) on the skins to ferment. If you want it to go faster, try adding a tablespoon of vinegar mother, but if you leave it for a week, you should have used up most of the alcohol and will have created your own tepache vinegar, which is less sweet when diluted (which I prefer). A careful 2022 study looked at the process of tepache fermentation and found that after seventy-two hours, around half of the original sugar content had been used by the microbes and at least ten types of yeast were found growing in the mix in addition to many LAB and AAB microbes. The analysis confirmed that initially the process starts with creating lactic acid and alcohol and then the microbes switch to producing acetic acid (vinegar) after around three days.

### *Vinegar in five*

1. Vinegar consumption (2–3 tablespoons per day) can modestly reduce blood glucose levels and possibly blood pressure, but it should be diluted before drinking.
2. Having a vinegary dressing on salad before a main course is a good habit to adopt.
3. The type of fermented vinegar you use probably doesn't matter too much for health but does for taste.
4. Tepache is a sweet fermented Mexican drink made from pineapple skins, which if left longer, turns to a tasty vinegar.
5. Vinegar-making is a great way of reducing food waste from leftover wine and fruit scraps such as apples (apple cider vinegar) and pineapples (tepache vinegar).

## Fermented soy

### How to make Kenji's simple miso

- To make 500g miso, you'll need 150g dried soybeans, 150g rice koji (bought online), 108g sea salt.
- Soak the beans overnight in cold water.
- Drain, rinse and add to a pan with fresh water, then simmer for 4–6 hours until soft.
- Drain the beans, saving some liquid. Mash the beans and liquid in a food processor to make a paste.
- Mix the rice koji and sea salt together, then add to the beans and blend together.
- Roll into balls and pack into a clean 500ml Kilner jar, with the lid pre-sprinkled with sea salt, add layers of plastic or cling film with weights to the top to keep it airtight on top of some extra salt.
- Store in cool, dark spot away for 6 months.

The soybean has been cultivated in China and probably Korea and Japan for over three millennia, even though in its raw state it is bitter and inedible to humans (but not cattle). Apart from boiled or steamed edamame, it requires a highly complex and staged system of drying, fermenting, brining, roasting and storing in jars and barrels to produce the huge range of products, like soya milk and tofu (its ricotta equivalent) or tempeh (its fermented cheese equivalent) or soft, smelly natto (a bit like Brie). The fermented bean also produces soy sauce and, as its waste product, miso paste. The ingenuity and patience of our ancestors in China, Korea and Japan to see the potential of this inedible raw bean was remarkable.

While I love to start a meal with salted or spicy edamame, which are merely steamed or boiled and not fermented, we'll focus here just on the fermented soy products that take longer to prepare. This means we won't look at soya milk or tofu, its high-protein by-product.

Soy is now a major US cash crop, used in cattle feed and in most ultra-processed foods in some form and is heavily subsidised by the US taxpayer. It is mainly used in low quality products: soy protein to bulk up a product's high-protein claims, soy oil as a source of polyunsaturates, soy lecithin as an emulsifier, to name a few examples. Most soybeans in the US have been genetically modified to be resistant to the herbicide glyphosate (known by its brand name Roundup) and there is concern about whether soybean consumption increases risks of some cancers. This is said to affect women especially, as they contain chemicals called isoflavones, which can mimic sex hormones. But I don't buy this: in East Asia increased soybean eating is actually protective against cancers, which suggests the soy–cancer association is more related to the intake of ultra-processed foods in the US than anything else. Most soybean research, although incomplete, suggests that the isoflavones in soy, acting on oestrogen receptors, generally have beneficial effects on metabolic problems and obesity.

Fermentation changes the structure and chemicals of soy radically and fermented products have not been associated with cancers. Reassuringly, over thirty-two clinical trials have compared eating plant proteins (mainly soy) vs animal proteins in patients with high cholesterol and found overall they improve blood fat profiles and so are beneficial for the heart. Another summary meta-analysis in 2023 looked at eleven trials in healthy subjects, randomising them to either animal protein or plant protein (again, mostly soy). They found similar results, with modest reductions in bad cholesterol (LDL) and reductions in mortality. So eating soy whole or its derived proteins, outside of ultra-processed foods, appears fine for health and may have some benefits, even if not fermented. The process of fermentation, however, introduces welcome complexity to the flavour and a great cocktail of health-supporting bioactive compounds.

## Fermented soybeans

The first fermentation of soybeans originated in China, where they were typically cooked, cooled, dried and mixed – originally by accident – with the koji fungus, then left in the high heat of summer to allow fermenting to transform the dull beans into something exciting. The fermented beans were then mashed into a paste and used to liven up many other foods and provide a long-lasting source of protein. Over later centuries, it is likely that dried fish and meat were added to the fermented bean paste to create the umami flavours beloved of East Asian cuisine. Soy sauce is probably its most famous offspring, as cooks started experimenting with making the bean pastes more liquid and then separating out the salty liquid as a condiment. Although China and other countries undoubtedly had earlier forms of soy sauces, Japan made the sauce a national icon, in the thousand-year period when meat or meat sauces were strictly forbidden under Buddhist emperors. Soy sauce then slowly filtered into Europe after the seventeenth century when trade with Japan restarted. The fermented beans are no longer as popular, but two spin-offs certainly are: miso paste and soy sauce.

The traditional way of making soy sauce is to boil the soybeans for around six hours until soft, then mix them with wheat and the koji fungus at a warm temperature with plenty of air until the koji mould produces spores. The beans are then placed in a very salty brine that kills off the koji, and a second fermentation takes place, lasting from six months to several years, traditionally in barrels. The liquid is then separated from the solid and sold as prized soy sauce. Now sadly most industrial soy sauce is no longer made with fermentation, but with chemicals under pressure that break down the beans using hydrochloric acid. Then other enzymes and flavourings and colourants are added to imitate the original at a cheaper price. Only about 1 per cent of soy sauce in Japan is still made traditionally, though you can still find industrially made but fermented brands like Kikkoman with live microbes. There are similar stories with Korean and Chinese versions, many of which are still fermented. Soy sauces that are lighter in colour are oddly usually saltier than the more common dark versions; the darker (less salty) versions usually contain more diversity of microbes. Wheat (gluten) free versions, called tamari, are increasingly popular; these are a mixture of fermented soybeans and rice.

### *Miso paste*

This soybean variant is so versatile that I now use it most days when cooking instead of traditional stock cubes. To make it the mashed beans are mixed with the koji fungus, initially in aerobic conditions. The salted mixture is then packed into fermentation containers and sealed to prevent oxygen exchange for a second, much longer, anaerobic fermentation. Lactic acid bacteria convert the sugars in the soybean mash into lactic acid, carbon dioxide and ethanol, helped by the yeast *Zygosaccharomyces rouxii*. Different salts, timings and ingredients all using koji as a starter will create very different products (rice miso, barley miso or soybean miso) and often a mature miso is added as an extra starter. Fermentation alters the colours and taste so white miso is sweeter but less complex than red miso, which is fermented for longer. Miso paste still has to be made naturally, but it lasts a long time if refrigerated and is usually not pasteurised. There are many variants of miso paste, both within Japan and many other countries, but they are all salty and have a strong umami flavour and because they are alive, they are kept in the fridge.

The most common use of miso is in miso soup; in Japan, unlike in the West, the soup is often eaten at the end of the meal. The miso paste should always be added at the end of the preparation after the other ingredients (dashi broth, bonito flakes, spring onions, etc.) have been added to water; this is to ensure that some of the microbes will survive, depending on the final temperature of the soup. So always avoid using boiling water for miso soup and ideally wait until it is below 50°C before adding the miso paste to help the five types of bacteria in the end product survive. These are: *Bacillus amyloliquefaciens*, *Bacillus subtilis*, *Staphylococcus kloosii*, *Staphylococcus gallinarum* and *Lactococcus* sp. GM005. The koji fungus will already have been killed off.

## *Natto*

This pungent, slimy version of fermented beans is made with a specific bacteria called *Bacillus subtilis natto*, which is found naturally in Japanese straw and is nowadays added as a starter to the beans. It breaks down the beans and produces a mucus-like gluey substance that envelops the beans and your chopsticks. It is a real delicacy in Japan, where they eat 700,000 tons of it a year and assume foreigners will find it revolting – and many do. If you have never encountered it before, it has been compared to eating a runny, smelly cheese. I was shocked when I first saw it at a hotel breakfast buffet but I have grown to like it, especially with runny eggs and some rice vinegar and spicy pickles (onsen style). There is absolutely no doubt it is alive and full of microbes (*Lactobacillus plantarum* and *Lactobacillus paracasei*). Unlike other acid-producing ferments, the *Bacillus* makes an alkaline environment, breaking down the glutamates in the bean, and defending itself that way.

Natto has many proposed health benefits and has been the subject of a surprising amount of scientific research, compared to British baked beans for example. Natto contains a range of essential nutrients and bioactive compounds. It is one hundred times higher in vitamin K2 than most cheeses. Its star bioactive chemical, produced by the *Bacillus natto* microbe, is an enzyme called nattokinase. This has been shown in lab and rat studies to have major effects in eliminating blood clots and reducing dementia. The Japanese have launched a range of natto and nattokinase supplements globally aimed at preventing heart disease and reducing blood pressure. If this catches on in the West or if they ever have any good human data, we could see natto, egg and chips being a new health food.

## *Tempeh*

Tempeh deserves a brief mention here too. It is a fermented product using the whole bean originating from Java in Indonesia, but it is usually marinated and fried or deep-fried, so most of the twenty-five plus microbes present are sadly destroyed by cooking. The whole beans are mixed with the spores of a filamentous fungus *Rhizopus* and left to ferment for twenty-four hours in heat. As it cools for the next twenty-four hours the mould dies off and the beans are meshed together with the slimy filaments of the fungus. Afterwards it is ready to be pressed into shape – the resulting cake is then stored or cooked.

With around 20 per cent protein, tempeh is very healthy; it also contains B vitamins and importantly vitamin B12, iron, calcium and manganese, which is particularly helpful for vegans. It can also be eaten raw to maximise the health benefits. I was sent a sample by a West country couple who started their company Tempeh Meades after learning the trade in Java. They make it by hand and ship it round the country and gave me some tips on eating the live version. But I found just a nibble of the raw tempeh hard going, so, like most people, I prefer to marinate it in some miso paste and mild chillies. I then pan-fry it, as I would halloumi cheese, and found it had great nutty, mushroom-like flavours. The fresh stuff only keeps for around forty-eight hours, so I put the rest in the freezer; it lasts for ages there and defrosts well.

## What to look for when buying fermented soy products

Soy sauce is a fantastic condiment with a tasty umami hit with plenty of antioxidants. It can be fermented or acidified and it is worth paying a bit more for the fermented versions to get a greater range of flavours. So do look at the labels, which should say something like 'naturally brewed'. Most of the popular Kikkomann range that I have seen are fermented.

For miso paste, if it's kept in the fridge and doesn't say it is pasteurised, there is a good chance that it still has probiotic activity, as well as all the helpful chemicals and the ability to fight off other pathogenic microbes and toxins. When you make miso soup, the paste rather than the dried powder is a better bet, but remember – if possible – to only add it to liquids below 50°C to keep it alive. The same dilemma occurs when using miso paste instead of a commercial stock cube, but you have to accept that often it is not practical; sometimes it is just about enjoying the flavours and hoping the dead bugs can still work.

Both soy and miso have high levels of salt, which could be a potential problem for blood pressure as we discussed for kimchi (see page 99). One Korean study explored the local version of miso called jang and compared high and low consumers in 58,701 men. What they found is that men with high salt intakes from non-jang sources had higher blood pressures and cardiac risk factors, but men with high jang intakes had lower risk factors and lower weight and blood pressures. So it appears that despite its high salt content, the other benefits of miso or jang, whether from its probiotic microbes or products, counteract any negative effects of the extra salt. As always these results are averages in a population and individual responses can differ.

For tempeh, freezing and cooking it will kill off the amazing range of twenty-four microbes, but fortunately not all the health benefits resulting from their hard work. The postbiotic products are still there after cooking and several studies show that pre-fermented tempeh has significantly higher levels of antioxidants and polyphenols (and isoflavones) than the simple boiled bean versions – this probably holds true for all soy versions. Finally, natto, unlike tempeh, must be eaten fresh, so while you can mix it with raw eggs and rice, don't try to microwave or cook it.

You can now buy miso paste in many shops and supermarkets. Making your own miso is not that difficult, but it does require time and patience. The simplest way is to buy some koji spores and dried soybeans on the internet and make big batches. My fermenting friend Kenji Morimoto showed me it wasn't that hard (see page 256). Although it will be months before you find out if it was a success or not, the wait is worth it when you finally taste your first homemade miso, as I found out.

### *Fermented soy in five*

1. Miso, tempeh, natto and soy sauce are all tasty, healthy products despite the salt content.
2. Most miso and natto products are alive with probiotics and need to be treated carefully.
3. Miso is a great substitute for stock cubes and can be used to ferment other vegetables, but it should never be heated over 50°C if you want the probiotic benefits.
4. Soy sauce is often pasteurised or unfermented so check the labels carefully.
5. All fermented soy foods have some health benefits even if the microbes are dead.

## NOT-SO-LIVE FERMENTS

### Sourdough

#### How to make simple sourdough rye bread

- Put at least 3 tablespoons of sourdough starter into a large bowl with about 100g of rye flour and 100ml of tepid water, mix and cover with a damp tea towel. Leave for 12–24 hours.
- When bubbling and lively, put 3 tablespoons of the new mother back into the fridge.
- Add 500g of flour (300g wholemeal rye and 200g wholegrain white or spelt) and 3 teaspoons of sea salt and mix until you have a spongy, mobile wet mass.
- Cover again with a wet tea towel and wait another 12–24 hours, occasionally folding and stretching it.
- Grease or spray a 900g loaf tin (or lidded Pyrex or cast-iron dish) with vegetable oil and shake on some flour to stop it sticking. The smaller the container the better the rise.
- Pour the dough into the dish, cover and let it rest and rise again for 1–2 hours.
- Put in an oven preheated to its highest setting (240°C/220°C fan) for 30 minutes, then carefully remove the lid, reduce the temperature to 200°C/180°C fan and bake for a further 30 minutes.
- Tip out and allow to cool for at least 60 minutes before slicing and enjoying.

Many delicious foods use fermentation to give them their distinct qualities, even if the subsequent cooking process kills off the microbes. But, as we have seen, microbes can still provide health benefits from beyond the grave, and perhaps the most celebrated example is sourdough bread.

There is a long history of fermenting grains with fresh yeast starters to make breads in many cultures around the world. This is how all leavened bread was made before the discovery of dried shelf-stable yeasts by Americans in the Second World War. Fermented bread is made by allowing dough to ferment with the help of bacteria and yeast, which break down the sugars and starches in the flour and create gases that get stuck in the structure of the proteins and cause the bread to rise. This bread is known for its distinct flavour of sourness, due to the production of lactic acids, and chewy texture, as well as its potential health benefits.

There are many different examples from around the world. Sourdough bread is made with a sourdough starter, which is a mixture of flour and water that has been allowed to ferment with the help of wild yeast and bacteria. Pumpernickel bread is a dense, dark bread that is made with rye flour and a sourdough starter. The bread is traditionally baked slowly at a low temperature for up to twenty-four hours, which gives it its distinctive flavour and texture. Naan bread, a staple flatbread in Indian and Pakistani cuisine, is made from a dough of flour, water and yogurt or milk, which is left to ferment for several hours before being cooked in a tandoor oven. The fermentation gives the bread a soft, chewy texture and a slightly sour flavour. Injera is a fermented flatbread staple in Ethiopian and Eritrean cuisine. It is made from a batter of teff flour and water that ferments for several days, which gives it a sour flavour and spongy texture when cooked like a large pancake.

Ever since Vanessa Kimbell of The Sourdough School taught me the rudiments, I love to make sourdough bread regularly and I can do it in ten minutes a day if I'm organised. It exploded in popularity in the early 2020s, thanks to a proliferation of small bakeries, and many of us trying to make it at home, especially during Covid lockdowns, leading to some shops running out of flour. The big manufacturers have also caught up and all supermarkets now sell some form of sourdough loaves, although the vast majority are what are called sour-faux loaves, being made industrially with chemicals and only adding a few grams of sourdough flour as a placebo plus yeast to allow the claims on the label. Sadly, food regulators have not helped consumers (apart from in France and Germany) to separate the fake from the real loaves, as there is no legal definition of sourdough and some manufacturers claim yeast doesn't have to be added to the ingredients list as it is part of the process.

A real sourdough should contain very few ingredients and no yeast or baking powder and as it takes longer to make it will likely cost more. I visited a sourdough bakery called Bertinet in London that has scaled up its operation to produce high-quality loaves around the country. It was amazing to see huge vats of starter (or mother) being manipulated using pretty much the same systems we use at home. They are now supplying major supermarkets and even have a sliced version, which tastes great.

Real sourdough is the antithesis to the modern Chorleywood white loaf that is made industrially in a couple of hours. It is not hard to make, but takes time, careful planning, good ingredients and plenty of natural microbes. A whole matrix of microorganisms are present in any natural sourdough starter including LAB and yeasts, which break down fibres, feast on sugars, produce healthy metabolites and enhance the vitamin content found in the grain itself.

The fermenting process starts by feeding up the mother with some extra flour and water and waiting 12–24 hours for it start producing bubbles, then you add the boosted starter to your main batch of flour, water and a bit of salt and wait 12–36 hours (depending on conditions) until it is bouncy and bubbling. It lasts well for several days and the amount and type of flour can be modified, as can the length of fermentation.

So what are sourdough starters, or mothers? Remember these start out as a simple mix of flour and water, left at room temperature as bait to encourage friendly bugs to settle and grow from the surrounding environment. From anecdotal reports and a few experiments, we know the microbes found in starters will often vary, with over fifty strains of LAB and twenty species of yeast, mainly *Saccharomyces* and *Candida*. A 2020 study of sixteen sourdough bakers using standardised methods showed that the microbes on their hands resembled to some extent the microbes in their starters, giving them some unique characteristics. It turned out it was a two-way exchange of bacteria and yeast from their hands to the flour. Different starters will create quite different breads in terms of taste, consistency and rising time.

More recently, a worldwide consortium formed to explore some of the big mysteries about sourdough. They asked for volunteers around the world to use standard starting conditions to make their wild yeast starters and then compared them. They used microbial genetic analysis to fully characterise 500 starters from four continents around the world, all sent by post to Boston. Their findings were surprising. Geographic location had nothing to do with the microbial composition of the mothers, nor did climate have much effect. In addition, a few microbes that we didn't consider were major players and acetic acid bacteria (AAB) played a major role in determining the speed of the rise of the dough as well as the distinctive vinegary taste.

The sourdough starter communities all had a consistent pattern, with a few species dominating, like the pair *L. plantarum* and *L. brevis* that work together, or if they weren't present, *L. sanfranciscensis* would dominate, especially in starters from bakeries that had nothing to do with San Francisco. *Saccharomyces cerevisiae* was the dominant yeast strain (which explains why it is commonly known as brewer's yeast), though there were often many minor ones. The study, unlike our human experiments with ZOE, only used 16S genotyping, not the whole genome, so they may be missing subtle variations at the strain level that could be influencing flavour but couldn't be detected. So it turns out that much of the variation and key microbial contribution comes from the microbes hanging around on the flour itself, another very important reason to pick your flour carefully.

## Sourdough and health

Sourdough bread has been touted as a cure-all for everything. While it is true that if you buy or make the real thing yourself you can be sure of avoiding unwanted chemicals in most supermarket breads, it is still not clear that sourdough bread – though it tastes better – has any clear advantages compared to a homemade chemical yeast bread made with comparable flour. Of all the staples I ate and liked, bread – whether brown, malted, seeded, wholewheat or white – was the one that consistently pushed my blood sugar levels up. I thought switching to sourdough would solve that, but it didn't. Yes the peaks were possibly slightly reduced but not enough to really make a big difference . . . until I discovered sourdough rye bread. If I don't overdo it and make sure I have with it some olive oil, cheese or avocado I can get away with it. So now making rye (or mainly rye/spelt) sourdough is part of my routine; I now find other bread hard to eat and very sugary.

But how much of my experience is supported by science? A 2023 review looked at the evidence from twenty-five small clinical trials of sourdough vs yeast breads. The total number of participants was 542, with fifteen of the studies looking at blood sugar control in healthy and unhealthy subjects. The results for reducing blood sugar were mixed, with only half showing a positive effect and the others no significant differences. For appetite the results were slightly better, with most showing sourdough kept people feeling full for longer, and similarly most of the studies looking at people with IBS showed a more convincing reduction in symptoms with sourdough, not seen in healthy individuals. No clear benefit on heart disease markers was seen in seven studies. But these were small studies, with short durations and a wide range of flours and fermentation methods being tested. There are so many variables in making sourdough bread, as well as the different flour possibilities, that it is hard to be sure of clear mechanisms. I was interested in several studies that suggested that fermentation of rye bread by microbes causes a sticky amylose layer to form, so reducing the speed at which the starch in the rye can be broken down to sugar in our guts (resistant starch). This may be partly the reason I get lower sugar spikes with my homemade rye sourdough. If you are unlike me and not sensitive to carbs or gluten, then choosing sourdough bread is a matter of taste and cost and knowing that you are buying a product made with care.

Most people with coeliac disease (CD) would be horrified at the thought of trying wheat bread again, but in line with the data of benefit on IBS, some small Italian studies found that four out of six sufferers could tolerate (under supervision) long-fermented sourdough bread. Microbes in sourdough produce proteolytic enzymes that help degrade the gluten into smaller fragments than in normal bread, which means these protein fragments, or gliadins, are less allergenic and so less likely to cause problems in those with CD or intolerance. Most people would expect sourdough bread to be better for their gut microbes compared to commercial breads, but a small Israeli study found no differences after a couple of weeks. While this could just be too small and short a study, they also found that the responses varied enormously from person to person, again emphasising the importance of personalisation in response to ferments due to the different make-up of our gut microbes.

Using high-quality flour is important if you have access to it. I like to use stoneground wholegrain flour, ideally organic so it is low in pesticides and high in fibre and packed with nutrients. The blends I prefer are a mixture of a small amount of wholegrain and milled wheat flour, which produces plenty of gluten and holds the gas pockets well, providing some sponginess; a greater amount of rye and/or spelt wholegrain flour, which are higher in fibre and nutrients and provide a denser, moist texture and nutty flavour; and finally supplemented with a nut, seed or plant mix to add diversity, texture and yet more fibre.

## *Sourdough in five*

1. Real sourdough bread is generally healthier than commercial breads with multiple ingredients.
2. However, sourdough is only as good as the quality of the flour, fibre and sugar content, as well as the diversity of plants added.
3. There is some weak evidence that sourdough may reduce blood sugar spikes compared to commercial breads and stronger evidence that it helps IBS symptoms.
4. High-fibre rye breads may be the healthiest option (for me at least).
5. It is fun and quite easy to make your own sourdough regularly.

# Tea and coffee

## A simple tea

- Source pre-fermented quality tea leaves (such as Darjeeling or lapsang souchong).
- Add 2 teaspoons to a pre-warmed pot.
- Add freshly boiled water at 90°C.
- Wait 3–4 minutes.
- Strain and serve black or with milk.

## Tea

No recipe is really needed for the most popular manmade drink in the world. The Chinese started to add leaves from the *Camellia sinensis* tree to boiling water over 4,500 years ago. Green, white, yellow, oolong, black (red) and dark tea are the traditional classifications based on the degree of fermentation, yet all are made from the same *C. sinensis* leaf. But what role do microbes play in making tea before we drink it? Not much, it was thought until recently. But we now know that the at least eight types of bacteria and fungi on the leaves play a key role, producing extra polyphenols and imparting special flavour characteristics.

Dark tea from China is the most thoroughly fermented tea, giving it many theoretical health benefits, but green tea, which is hardly fermented, has many more health claims. Most teas have anti-inflammatory and antioxidant effects in test-tube and animal studies, but the medicinal claims made for some teas far outweigh the scientific evidence.

### *Black tea*

For classical teas, just two or three of the youngest leaves from the bush are picked and 'nipped in the bud', a process which is still best performed manually. As soon as the leaves are plucked, the chemical composition starts to change: the polyphenols leak out of the cells and are oxidised by enzymes. If allowed to dry naturally in the sun, this process turns the leaves black. Once dried and pressed, this becomes black tea. Oolong tea is made from young leaves and fermented for a shorter time. The fermentation takes place naturally as the leaves are laid out on withering racks and repeatedly rolled.

Black tea is heavily fermented and usually contains around twice the caffeine content of green tea, but the amounts can vary considerably. During microbial fermentation, several unique compounds associated with dark tea quality, taste and scent are formed, such as catechins, flavonols, flavones, phenolic acids, alkaloids and terpenoids, thanks to the work of fungi including *Aspergillus* and *Candida,* and bacteria like *Bacillus*, *Pseudomonas* and *Brevibacterium*. Black teas contain over 600 volatile chemicals which, in combination, can produce fruity flavours.

My favourite daily brew is Earl Grey tea, a black tea flavoured with oil of bergamot, which I was disappointed to hear was originally added to mask shipments of poor-quality tea. Unfortunately, I can't use Earl Grey to make kombucha, as the bergamot oil interferes with the microbes. Some black teas, like lapsang souchong, are smoked while drying to give strong distinctive flavours.

A strong black tea brew with a splash of milk (now usually made with tea bags of Assam and Sri Lankan teas and known colloquially as builders' tea) is still the drink of choice in the UK, Ireland, Australia and New Zealand, though it is dropping rapidly in popularity. It is slightly different to English Breakfast tea, which is lighter and often made with Kenyan leaves.

## *Green tea*

After a few hours of steaming or roasting, the oxidation/fermentation process is halted. After drying, this becomes green tea. These leaves retain most of the properties of the fresh leaf, especially the levels of polyphenols, which are higher than in black tea, but less of the complexity that comes from fermentation. Japanese teas are mostly green teas and vary by region, the soil they are grown in and the precise time the *Camellia* leaves are left in the shade before picking. The shading process increases the chlorophyll content, polyphenols, caffeine and theanine, which (apparently) provides the calming Zen effect. Jasmine and other scented teas are made by mixing the flower in a jar with the green leaves and leaving for up to twelve hours. Pu-erh tea is made by rolling and slowly drying large green tea leaves. Then, as it is packed together, it allows gentle fermentation to occur, often driven by the fungus *Aspergillus niger*. The leaves are bundled together, pressed into a cake and left for around three years or more, to continue fermenting, slowly increasing the complexity of the aromas. As well as the common fungi, hundreds of other species have been found, including some fungal toxins, which is why it is Chinese custom to throw away the first toxic brew, before enjoying the safer second brew.

Matcha is now popular as a sugary, milky latte, but in its original form is the tea used in ceremonies. It is a concentrated, ground-up powder of the shaded and early-picked young shoots, which is made into the classic tea by adding water and whisking it for a few minutes to produce a froth. As a whole-leaf ground powder it has one of the highest fibre contents of teas, at around 17 per cent, which despite its lack of fermentation may play a part in its miraculous-sounding health claims. One tablespoon has around 6–7g of fibre. It has three times more polyphenols than green tea including catechins and gallic acid. As it is essentially concentrated green tea it is likely to have even more health benefits than regular green tea, as long as it is not smothered in milk and sugar. Another that I particularly like is Genmaicha, which is a blend of green tea with roasted rice, giving a complex, nutty flavour.

## Tea and health

Tea was originally used in China for medicinal purposes and it has kept its reputation as 'the healthy drink'. It was originally thought to reduce cancer, although the latest data fails to support that. For reducing heart disease there was a modest benefit seen in summarising thirty-five studies on drinking 1–3 cups of black or green tea per day, showing 10–15 per cent lower rates of heart disease and stroke. But if you drank 4 cups or more of black tea your risk started to increase, while for green tea, the preventive benefits increased with dose. Green tea also oddly showed no clear benefits in non-Asian populations. The difference in risk between black and green tea might be explained by the reduced polyphenol content in black tea, or simply by biases in the studies of culture and affluence. The fact that most people drink black tea with sugar and often milk, which inhibits polyphenol absorption, while green tea is usually unsweetened and without milk, are extra factors that are hard to account for without large clinical trials.

Green tea chemicals called catechins have been associated with theoretical benefits, such as reducing cancer risk, enhancing chemotherapy and aiding weight loss, dementia and Parkinson's disease, but the evidence is weak and based mainly on test-tube studies. Strangely, the European Food Safety Authority (EFSA) actually considers too much green tea, especially as artificial extracts, to be potentially harmful in high doses as it can damage the liver. Most teas contain some caffeine, although all are lower than coffee: black tea is typically 50–60mg per cup and green tea 20–30mg. Matcha is the highest at 70–90mg but also contains an amino acid called L-theanine that counteracts the caffeine; in clinical trials matcha was shown to have a mild anti-stress effect after a few hours and slightly improve sleep. A double-blind Japanese study versus placebo showed that taking 3g a day for twelve weeks could possibly improve cognition in elderly women.

Iced tea has been around for centuries, and in the Southern United States, three-quarters of tea drunk is now iced. The big tea companies have teamed up with the big soda companies to make incredibly weak iced tea, with the market leader (Liptons/Pepsico) only containing 0.1 per cent black tea extract. These commercial iced teas have no health benefits as they contain over 24g (6 teaspoons) of sugar (or high-fructose corn syrup in the US) per can.

One long-standing debate focuses on the quality gap between loose-leaf and tea bags. These bags took off despite the poorer quality of the brew, due to the lack of space the leaves have to infuse the water. Tetley introduced the square tea bag to the UK in 1953. While clear differences exist between leaf and bag, technology has improved over time with some innovative designs to increase the infusion area.

Whatever your choice of tea, it does appear to be generally beneficial for your health, unless you are worried about pesticides. Levels of herbicide chemicals like glyphosate (Roundup) are commonly found at detectable levels on tea leaves as most producers spray their crops extensively. Thankfully, studies have found that in brewing tea, only a small amount of common pesticides reach the liquid.

Fermenting your own tea is not that easy unless you decide to make bricks of pu-erh tea or – much simpler – you add some brewed tea with sugar to a kombucha mother (see page 246).

## Coffee

Coffee is another drink whose microbial fermentation is often forgotten. I used to assume it was an unhealthy option compared to tea and tried to cut down my consumption. But I now consider it a healthy drink – it's certainly much better for us than orange juice. Coffee appears to have a unique relationship with our gut microbes. When we looked at the gut microbiomes of 28,000 people using the ZOE database, we saw that of all the foods and drinks consumed, coffee had the strongest link with changes in our microbes. In fact, just by measuring one single microbe, we could tell if you drank coffee or not. This little-known microbe called *Lawsonibacter* may be the key to why coffee can prolong your life.

Fermenting coffee beans probably started in Ethiopia in the sixth century and when it finally arrived in Europe this exotic drink had a huge impact – it was the first time most Europeans had tried a stimulant drug, which, in contrast with alcohol, could sharpen the mind and keep them awake. Today, the majority of the world's population drinks coffee daily, with the Finns consuming the most per head at around 500ml, three times as much as Americans and four times as much as Britons.

Two main beans dominate the world market. The *Arabica* plant is the original, more delicate plant, which is more expensive to grow and trickier to keep alive; the other is called *Robusta*, which is hardier and cheaper to farm. Once picked, the green, ripe coffee fruits are separated from their seeds (beans). Microbial fermentation occurs after piling them up in hot moist temperatures and then drying them, or by using water to crudely clean them before letting them ferment for a few days – called dry or wet fermenting, respectively. Although there can be more than 1,000 species of bacteria and fungi in coffee ferments, studies show that the microbial community is dominated by the lactic acid bacteria (LAB) *Leuconostoc,* the yeast *Kazachstania* and the acetic acid bacteria (AAB) *Acetobacter*. Importantly, studies show that the quality of coffee is correlated with the number of different species, especially fungi and specific *Lactobacilli*, *Pseudomonas* and *Pichia* bacteria. Some coffee farms now add special yeasts and LAB bacteria as starters during their fermentation to improve flavour. Coffee is usually wild fermented in the open air, so any random yeast or bacteria can join.

The issue with long, sun-drenched fermentation is that it degrades polyphenols. To create an optimal environment for the microbes and remove the fruit pulp in just a couple of days, farmers are now using a 'backslopping' method learned from other ferments. This involves using some leftover fermenting water from the day before to start the next batch.

The dry and still tasteless beans are then roasted for a few minutes at around 200°C for a light roast and higher and longer for a darker roast, which puffs them up and brings out more volatile chemicals and distinctive coffee aromas while killing off any microbes. The hotter and darker the roast, the less polyphenols are contained in the coffee, so you're best off drinking a light-roasted arabica for polyphenol content, but not necessarily for flavour. With coffee, there is always a trade-off.

After roasting, the beans are cooled, rapidly locking in pockets of carbon dioxide, which protects them from going off. The intensity of the roasting determines the flavours and characteristics of the bean. Lighter roasts produce a fruity tartness, whereas darker roasts have more aromatic compounds, but if these are overdone they can destroy more subtle flavours, which is often deliberate if the beans are of poor quality. On roasting polyphenol content initially increases then reduces after about 9 minutes, with big variations between beans.

Of the common brews, espresso is made after 30 seconds of bean and water contact and has a 1:3 coffee-to-water ratio. This has the most intense flavour and produces a distinctive 'crema', formed by a mix of all the contents of the bean as they are forced out, including the polyphenols, fats, proteins, carbs and carbon dioxide. One shot of espresso actually contains less caffeine than most filter coffee servings, but because of the complexity of the process, it is highly variable and unpredictable. One study visited thirty cafes in Glasgow and found a three- to fourfold variation in caffeine levels of a standard shot.

Caffeine acts by blocking the action of a brain chemical called adenosine, which normally relaxes you and makes you sleepy, thereby increasing alertness and concentration. This may partly explain the increasing evidence that regularly drinking caffeine may delay the onset of dementia and Parkinson's disease.

What about decaffeinated coffee? I was once laughed at when I asked for a decaf espresso at Bar Italia in Soho late at night by a purist barista (who refused to serve me). Things have changed and now about one in five millennials and over 10 per cent of the UK market prefer decaf coffee. It shouldn't be dismissed as it is still a fermented product and probably provides the majority of the health benefits of full-caff.

Instant coffee also gets a bad rap and although I have grown out of it since my student days, it is basically industrially dried, fermented roasted coffee – and is at least a fifth of the price. Instant is usually made using the same initial steps as brewed coffee by freeze-drying or spray-drying the roasted beans after they have been mixed with water. Green beans are often re-added to provide some of the polyphenols like chlorogenic acid that would otherwise be low.

Whether instant coffee is as good for you as ground coffee is less clear. Large epidemiology studies have found no major differences in mortality between the types, but one study using genetic markers of coffee drinking showed that instant coffee users were not protected against type 2 diabetes and had increased markers of biological age. So, does this mean instant coffee with its cheaper industrial preparation is bad for us? Not necessarily, as it is almost exclusively made from *Robusta* and still contains around 80 per cent of the polyphenol content, although the addition of milk to make it palatable can reduce the antioxidant potential by about a third. Overall, instant coffee seems to be a cheaper but still healthy alternative to brewed coffee, if you go easy on the sugar and milk. Future coffee makers will need to encourage more science as coffee is increasingly viewed as a health beverage.

## Coffee and health

Coffee is not, as widely thought, a true diuretic. It does not dehydrate you, and it may have some benefits in reducing incontinence, though drinking over 3 cups a day can irritate your bladder and make you want to pee more. We also tend to forget that coffee starts life as a berry and is a reasonable source of fibre. When I first posted this nugget on social media, many nutritionists were critical because of our false preconception that coffee is not a healthy drink. But it turns out that for many Americans, coffee is their major source of fibre; each cup provides around 0.5g – 2 cups of Arabica contain roughly the same fibre as a small banana.

Coffee is generally fermented and digested twice, once in the wild by microbes, then once inside our colon, but sometimes there is a third digestion. Asian palm civets are used in captivity to produce kopi luwak. The coffee berries they eat are collected in their droppings and sold at huge mark-ups. These expensive coffees have no proven benefit and studies show that around 40 per cent are fake, and as captive animals are used, it is best avoided.

Many people (including my wife) find that they can time their bowel movements to the minute after their morning coffee; for me it is not so clear-cut. The power of a cup of coffee is well known to gut surgeons. Thirty-two trials have shown that giving coffee to patients after bowel surgery speeds up their first stool by fourteen hours, allowing them to go home a day and a half earlier than tea-drinkers. It also works in severely constipated children, where regular coffee had powerful results, but even decaf coffee had mild stimulatory effects. Strangely, though we don't know the exact mechanism, coffee is also good for liver disease (both fatty liver and alcoholic cirrhosis) and is part of a recommended treatment plan. All these studies show individual variation, so the effects are quite personalised.

So, not all the effects of coffee are caffeine-related. From our work studying the specialist and newly discovered microbe *Lawsonibacter* using ZOE data, we know that when it ferments coffee, it produces the helpful short chain fatty acid butyrate. We have worked out that the microbe also eats the caffeic and quinic acids in coffee and turns them into other interesting chemicals. While coffee drinkers have 5–6 times the levels of *Lawsonibacter* as non-drinkers, most non-drinkers have some of the bacteria hanging around eating scraps of other foods, and women strangely have more than men.

This is very much a modern microbe as babies and toddlers don't have it, nor do ancient samples and tribes who were never exposed to coffee. Coffee drinkers presumably transfer the coffee microbes to other non-drinkers in the household at low levels. How it got into our gut in the first place remains a mystery, but maybe it evolved slowly from eating something else once coffee came along. To get all the benefits of coffee, try a doubly fermented version called coffee kombucha, which you can rarely find commercially, but is easy to make yourself.

## *Tea and coffee in five*

1. All teas made from fermenting the *Camellia* leaf (green, black, dark, pu-erh, oolong) contain large numbers of polyphenols and antioxidants and are healthy.
2. Green tea is less fermented than black tea but has more evidence for heart benefits when drinking more than 3 cups per day.
3. Drinking up to five cups of coffee a day is even more beneficial for heart disease and possibly cancer and dementia; coffee also contains moderate amounts of fibre that help constipation.
4. Drinking decaffeinated or instant versions of coffee appears to confer similar health benefits.
5. Drinking tea or coffee with milk will reduce absorption of polyphenols by around 30 per cent and so is less healthy.

## Wine

### How to make simple wine (good at best for punch/sangria)

- You will need a very clean 5-litre glass container with a narrow neck (demijohn).
- To this add 1 can of concentrated grape juice (or other fruits), 1 packet of wine yeast and 4 cups of unrefined sugar.
- Mix well and top up with 4 litres of boiled and cooled water.
- Place a large uninflated balloon over the neck of the container and then place in a dark spot that is 15–20°C for roughly 6 weeks.
- The balloon will start to inflate after a day – when the balloon deflates, filter into bottles and drink.

Fermenting alcohol has been part of human history for millennia, helping us socialise, forget our problems and sometimes cause mayhem. We don't normally think about the healthy aspects of drinking, but drinks made with fermented grains or grapes contain valuable antioxidant chemicals in higher amounts than found in the basic grain or fruit, as the alcohol reacts with the microbes to produce hundreds of new chemicals. Red wine classically has the highest concentrations of different detectable polyphenols and nearly three times that of white wine, although a glass of apple cider has similar if not slightly more polyphenols than the average glass of red wine. Spirits, although they are also products of yeast fermentation, have the least nutrients of value: being over 20 per cent alcohol, no bacteria or yeasts can survive and any helpful chemicals produced during fermentation are destroyed by distillation.

It's no secret that I like a glass of wine. In the past, whenever I have recommended a glass of red wine as being good for gut health, I have gotten into trouble from earnest colleagues who believe that no type or amount of alcohol is good for you (and shouldn't be recommended), or from abstainers who have had previous problems with alcohol. But while realising the risks, there are clear differences between alcohols. Distilled spirits like whisky, gin, cognac, rum or vodka contain virtually no microbes or polyphenols so will affect our bodies differently.

During my time living in Barcelona, I noticed how the same older ladies would always meet at the end of the day for a glass of wine or beer in the evening sunshine, catching up and discussing their day. This social approach to drinking, where one glass of a favoured drink among friends creates a moment for social connection in the day, might hold the key to why some of the healthiest ageing populations in the world enjoy wine as part of their daily ritual. In 2021, one French agricultural minister – to help flagging wine sales and low production – went as far as to say that wine should not be considered in the same category as other alcohols, as it has special attributes. Why would wine be special? Well, the grapes and microbes have a lot to do with it.

We drink over seven trillion gallons of wine per year, globally. Although consumption is dropping, the French are still one of the biggest consumers, not counting those thirsty priests in the Vatican who drink twice as much as the average Italian because of onerous ceremonial masses. Grapes come in many shapes and colours. 'White' grapes are actually green or yellow in colour and are derived from the purple/red grape but genetically mutated to turn off production of anthocyanins, which are responsible for the red colour. White wine and orange wine are usually made with green grapes and red and rosé wines with red grapes. The key difference is that white wines are fermented without the grape skins, and red wines with the skins, while rosé and orange wines leave the skins in briefly.

The basic wine-making process has several stages. First, the grapes are collected and crushed (and for white wine pressed to remove the seeds and skin) and the sweet grape juice (called the must) is collected in vats to be fermented by yeast into wine. Sulphur dioxide (producing sulphites) and extra sugar or acid are added at this stage in virtually all wines, although yeasts can produce them naturally. The natural yeasts chosen are different strains of *Saccharomyces* that convert sugar to alcohol and carbon dioxide, but organic wine makers can start the process off with several organic yeasts found on the grape skins or in the air of the winery. Strangely, yeasts don't consume the alcohol they produce, they only use it as a defence mechanism. Another fermentation step called malolactic fermentation occurs in most red wines and some whites before bottling. Substances called fining agents are then added; these stick to the particles floating in the wine after fermentation to sink them to the bottom as sediment, keeping the wine clear.

Once the wine is clear, it is stored in barrels or large industrial tanks for months or years before a final filtering. The new wooden barrels impart extra flavour molecules to the wine, such as vanillin and woody lactones. At this stage the number of flavour molecules and polyphenols has increased dramatically, thanks to all the chemicals mixing with each other and being episodically exposed to oxygen. The wine is often filtered, then given an extra dose of sulphur dioxide to produce more sulphites, to preserve it before blending. It is then poured into bottles or massive tanks for transport and local bottling with corks (or increasingly rubber stoppers or screw caps).

Many people comment anecdotally that red wine may cause reflux or heartburn, but these are generally less acidic (0.55 per cent acetic acid) compared to most white wines (0.85 per cent, which is lightweight compared to vinegars of 5 per cent).

## Microbes in wine

Before fermentation it is hard to detect any big taste differences between the grape juices, but once the microbes have transformed it, after just a few days there will be a huge range of notes and flavours. The microbiome of any wine vineyard is dependent on the geographical location, grape variety, climatic conditions and agronomical practices, establishing a new variable of the concept of *terroir*. We now know the microbial profile of grapes can predict which key wine metabolites will give it flavour. The *Saccharomyces* yeast is still the main player, creating alcohol and a range of chemicals that provide the basic aroma matrix of the wine, but at least fourteen other yeast species have been shown to be involved in aromas. But we know the role of lactic acid bacteria (LAB) is also key. They are responsible for changing malic acid to lactic acid in secondary malolactic fermentation, required for most red wines and some white wines to make them more palatable and buttery. *Oenococcus* and *L. plantarum* can drive this process and are often present in starter cultures. Apart from the role of wood during the ageing process in barrels, other microbial-derived compounds also contribute to the flavour of wine. These come from the debris of dead yeasts and bacteria cells (called the lees) at the bottom of barrels and bottles. This is why the intentional use of lees sediment during wine ageing is so important.

Wine made from grapes is one of the most complex foods we ingest, containing thousands of chemicals, many of them the gut-friendly antioxidant polyphenols from the original grape berry multiplied by the fermentation process of wine-making. Stare at the red liquid in the glass and have a sniff and then a sip – 'mmm nice' you say. Most people stop there and miss much of the hard work of the microbes. But try harder, have a longer look, inhale and swish the liquid around all parts of your mouth for a few seconds, mixing it with air before exhaling and swallowing.

If you focus on your tongue you can feel the astringency due to the tannic grape polyphenols that can provide a guide to age and potential. If the colour has faded somewhat and is more a brownish purple this will confirm it's an old wine. From the heavy streaks of wine that trickle down the glass you can tell that it is high in alcohol and comes from a sunny vineyard. Studies have shown that different microbes inside our salivary microbiome play a key role here. Just like our gut microbiome, our saliva is a complex ecosystem of microbes that varies widely between us. Because of this, experiments have shown we all taste wine slightly differently. Some microbes can switch on chemical receptors in our mouth and tongue so we taste more of a particular compound (citrus, for example), while other compounds can actually break down the wine molecules rapidly and via chemicals (such as glycoside and cysteine conjugates) can generate novel aroma compounds. So although it appears pretentious, it turns out if we spend time swilling wine around our mouths, we are literally refermenting our wine in a unique way and liberating new chemicals before it reaches our guts.

## Wine and health

Most epidemiology studies show any amount of alcohol is harmful. Buried in the weeds of the research is that drinking around one unit per day seems to be the sweet spot for reducing heart attacks, possibly via a reduction in clots for blood vessels, though the risk of many other diseases, including cancers, is slightly increased. Although large studies claim no adverse effects of drinking less than seventeen units weekly on your biological age, other large studies claim there is no safe minimum level for overall mortality. My view is if you already drink alcohol and can drink in moderation without becoming addicted, your best health option (if you enjoy it) is red wine. I say this in spite of the fact that, if you drink one glass a day, the epidemiology modelling studies show your risk of a bad health outcome is increased, but this risk is statistically very, very low. Driving a car regularly is probably much riskier, so as usual it is important to put risks into perspective.

A 2023 meta-analysis review of twenty-five observational studies that explored wine drinking and heart disease concluded that there was a small but significant decrease of heart disease risk (and death) of around a quarter in wine drinkers, and it was similar in men and women, although the data was not good enough to assess quantities or the colour or types of wine. Randomised controlled trials are generally believed to provide higher quality evidence than observational epidemiology; however, for alcohol, these are often far too short to show health risks, as giving volunteers booze or placebo drinks in the name of science long term is tricky. A 2023 review looked at ninety-one small clinical studies of the effects of red wine, seven of which lasted over six months. They found consistent modest effects on reducing blood markers of oxidation and inflammation, and in the longer term on reducing markers of kidney disease and heart disease in patients with diabetes. The same review also looked at five short-term red wine and gut microbiome studies with 10–52 subjects studied for up to twelve weeks; three out of the five showed an improvement in microbiome health. One of these was a Spanish study that looked at ten people in a randomised, crossover, controlled intervention study. Over four weeks they found that the daily consumption of two glasses of red wine, red grape juice with 0.4 per cent alcohol, or two shots of gin all slightly improved the microbiome diversity profile. The two grape drinks (but not the gin) improved beneficial microbes like *Prevotella* and *bifidobacteria* and also reduced lipid levels and blood pressure, with effects being greatest for the fermented red wine, suggesting the complex chemicals formed by fermentation have a benefit greater than grapes plus alcohol separately.

A more recent and more complete trial was on forty-five men given two glasses of red wine per day for three weeks. The study showed significant shifts in the gut microbiome diversity and an increase in *Prevotella* and *Ruminococcus* species. Analysis of the chemical metabolites showed increases in the antioxidant chemicals in the blood. To confirm the small clinical trials my group performed a large, international observational study of 3,000 subjects including UK twins and Belgian and US cohorts. We found spirit drinkers – even moderate ones – have worse gut diversity scores, while red wine drinkers had the healthiest microbes, with white wine drinkers having a slight benefit, and beer drinkers neutral. There was a dose-response effect, with any benefit reducing dramatically at three glasses or more a day.

We are beginning to understand why wine could have this modest benefit that counteracts the negative effects of alcohol. It has a great variety of polyphenol compounds such as resveratrol, catechin, epicatechin, quercetin, and anthocyanin, which as well as altering gut microbe composition are also antioxidants and can change lipid profiles and reduce insulin resistance. Our own observational data from our Twins study and other cohorts suggests the importance of picking your healthy alcohol based on the number of complex polyphenols as well as taste. Based on this you would always pick red wine over white wine; there are a few long-term clinical studies comparing them, but a study of 157 Brazilians followed them for a year and found no clear differences in lipid levels between red and white drinkers, hinting that other factors are important.

## Chateau resveratrol

One ingredient in wine has reached celebrity status. Resveratrol, a polyphenol found in about seventy plants including peanuts, rhubarb and berries, but especially in red grapes, hit the headlines in the 1990s. This stuff apparently does amazing things to rodents, such as reducing inflammation, weight loss, helping brain flow, reducing cancer and even increasing lifespan. It also reduces blood pressure and improves heart muscle function, reducing the risk of heart failure. The antihypertensive effects are thought to be mediated via activation of two key cell pathways called SIRT-1 and AMPK, which result in nitric oxide (NO) production by cells in our blood vessel walls. These relax our smooth muscle and dilate blood vessels, thereby lowering blood pressure. Entrepreneurs started pushing wine because of these animal studies, although it was thought you had to drink about thirty bottles a day to get the supposedly beneficial resveratrol doses (400–1000mg/day).

I was very sceptical of the resveratrol story, particularly as it was based mainly on test-tube and animal studies and the massive revenues for the supplement industry. I also thought that even with training, I could never drink enough red wine. But I have recently changed my mind. While researching this book my eye was drawn to a 2023 randomised study that showed a clear human effect of resveratrol on the heart for the first time. The authors gave eighty patients with high blood pressure and at risk of heart failure, 400mg of resveratrol or placebo and found that after six months their heart function had improved significantly. There have been a number of other clinical trials, mostly showing reductions in blood lipids. But strangely, others given high doses (1,000mg/day) made blood lipids worse. A meta-analysis in 2022 of seventeen studies helped clear up some of the confusion and suggested they are overall likely to have a beneficial effect on lowering blood lipids (triglycerides and LDL cholesterol). But the dose of resveratrol appears to be crucial.

Several studies using doses above 400mg/day either showed no effect or clear adverse effects on lipids, while some other studies showed clear benefits as low as 5mg/day. A Spanish study gave twenty-five volunteers either two glasses a day of a low resveratrol red wine (1mg/litre), equal to 0.3mg/day, or a high concentration wine (5.9mg/litre), equivalent to 2mg/day for four weeks. In the 2mg/day group they saw an elevated blood protein that binds oestrogen and testosterone (SHBG) and is associated with less heart disease, reduced ageing and lower cholesterol levels. The results were much clearer for women than men but again show the potential of low doses and finding the sweet spot.

A summary of twenty-eight small human studies has suggested that while low doses of resveratrol supplements taken for over three months can slightly reduce weight (by less than a kilogram), high doses of 500mg or more can – in contrast – increase body weight. One study worked out that 5mg of resveratrol daily was the optimum dose for reducing cancer in the colon in humans and 1000mg was actually harmful. The potential of high doses to cause harm for substances we need only tiny amounts of and naturally get from our diet is not new; there are many examples where excess supplementation is harmful, such as calcium, vitamin D, folate and vitamin A. What I found most interesting is that the 5mg dose that does appear safe is roughly the same as you would get from a large glass of red wine.

In large Spanish population surveys, 98.4 per cent of resveratrol comes from wine drinking, mostly red wines, with a tiny amount from rosé and white wines. But resveratrol concentrations can vary widely across wines by around twentyfold, from nearly undetectable levels to 27mg/litre. I no longer try to drink thirty bottles a day to get the right resveratrol levels as lower doses will be more beneficial. If I picked the right high-strength resveratrol wine, I might get away with 1–2 glasses a day to meet all my needs. According to a Cornell study, it looks like pinot noir wines, with their thin-skinned grapes and high polyphenol levels to protect them from insects, have consistently high levels above 10mg; many New York state pinots produce over 13mg. But a more recent study of sixteen Australian and Kiwi wines from 2010–12 vintages highlighted even higher levels. Pinots and merlots scored well, from 4–27mg/litre, while the top wine, a Grampians shiraz, had a winning score of 26.8mg when tested three years after harvesting. But sadly two years later these levels had reduced by 75 per cent due to ageing and conversion of resveratrol to less active forms. So until resveratrol levels are printed on labels, it's best to drink your pinot and shiraz wines fairly young; at those high levels, just one glass will probably give you the magic 5mg.

Of course while it is fun, I am being reductionist in talking about just one of the polyphenol chemicals in wine, as there are hundreds more that could be more important individually or in combination. Polyphenol counts are generally highest in red wine that is aged in oak casks and uses cork stoppers. We don't know yet which polyphenols are crucial, but marketing your wine as having healthy properties is now big business – and is going to get bigger. In the meantime be wary of anyone selling high-dose resveratrol supplements: until we have more human data, low doses are always safer and should ideally come from natural food and drink.

## The science of hangovers

Ever woken up on a Sunday morning and wished you hadn't? The headache was bad enough, but the nausea and lethargy made you feel like you had been poisoned.

Hangovers are due to the side effects of our blood alcohol concentrations finally returning to zero after a night on the booze. The chemical produced when alcohol is broken down (acetaldehyde) is a poisonous substance that is degraded at varying speeds by different people; the longer acetaldehyde hangs around, the worse you feel. Generally, the purer the distilled liquid is to alcohol, the fewer congeners (chemical by-products of alcohol fermentation) there are and the lower the hangover risk. Vodka and gin are distilled to 95 per cent; malt whisky, cognac, rum and grappa are 70 per cent; bourbon 65 per cent; and tequila (my nemesis) only 55 per cent. The rates at which your body and brain return to normal are extremely variable. Studies have shown these rates are partly dependent on your age, gut microbes and genes.

Can ferments help us prevent hangovers? Several studies, mainly in mice, have shown the effects of probiotics found in cheese or yogurt (*Lactobacillus* or *Bifidobacteria)* or fermented ginseng in breaking down alcohol and counteracting alcoholic liver damage. For two weeks, fifty-four Koreans were given either a placebo or a probiotic mixture of four bugs (*Lactobacilli and Bifidobacteria*) that have enzymes that break down alcohol and aldehyde. They were then given vodka on an empty stomach. In those with western alcohol genes, the probiotics significantly reduced levels of both alcohol and its by-product acetaldehyde, but sadly not the self-reported symptoms, maybe because doses were too low. These mouse and human studies offer hope that a probiotic supplement mix or fermented cheese or a kefir shot could help mitigate human folly but only before, not after the event.

I tested my alcohol knowledge recently on a visit to the main medical school in Almaty, Kazakhstan, where after forty-five minutes of chatting and drinking wine, the dinner started with vodka toasts. I was seated next to the small, elderly dean, with my personal vodka waiter hovering by my shoulder. I failed to make my excuses and six double shots later, I wasn't so cocky. Just before my legs began to fail, I mumbled about having to meet a friend and staggered out. I next remember waking up fully dressed on my hotel bed at six in the morning, feeling (correctly) like I had been poisoned. I had hoped to sleep it off but had a prestigious lecture and prize-giving at 9.30 a.m. in front of 1,000 keen medical students. I was desperate for a cure. The hotel suggested a local remedy of fizzy horse milk kefir that only produced more nausea. I'll never forget the awful day that followed or the smiling face of the diminutive dean, looking immaculate and sharp as a button as he shook my shaky, sweaty hand and asked if I slept well.

## Wine headache

Around 10 per cent of the population complain of a specific wine headache, commonly blamed on sulphites that are formed mainly from the sulphur dioxide added to preserve the wine, and they typically and unfairly blame red wine. White wine usually contains more sulphites than red, because the tannins in the red wine act as natural antioxidants and preservatives, meaning fewer sulphites are needed than for white. There are actually about ten times more sulphites in dried fruit like apricots than in wine, yet we don't hear of many 'morning after' apricot headaches. A Portuguese study gave fifty-one self-reported sufferers two glasses of high- and low-sulphite white wine; only 27 per cent had classic symptoms that resembled a migraine headache with the high dose, and 14 per cent with the low dose. Of the control low-risk population, 20 per cent reported a headache after the high dose, showing only a small difference between the groups, and more people are sensitive to the suggestion rather than the sulphite. This supports an estimate by the US FDA that around 1 per cent of people are affected, a figure that is likely higher in Asian populations. It is virtually impossible for someone who likes wine but gets the occasional bad headache to completely avoid sulphites. Generally, the sweeter the wine, the more they are added as preservatives, and even organic or natural ones contain them, though usually at lower levels.

A few people get a very specific form of headache called red wine syndrome where they feel sick and go red in the face about 15 minutes after drinking a glass. Higher levels of compounds called biogenic amines (such as histamines in red wine) have been blamed, although one study tested this in self-reported sufferers and failed to show red wine was any worse. Other potential headache culprits that have been blamed (inconclusively) include the high tannins in some wines such as those from Bordeaux and Italian Barolos that may increase the brain chemical serotonin, and even the microbes formed by the second malolactic fermentation performed in many modern wines.

## Waiter, my wine has trichloroanisole (TCA) in it

I used to mistakenly think corked wine was caused by the destruction of the cork, so letting in oxygen, and *Acetobacter* bacteria turning it to vinegar. Unfortunately corks can attract moulds and bacteria, which produce their own unwanted aromas. These can be earthy like mushrooms or smelling of damp basements, or sometimes reminiscent of the aroma of wet dogs splashed with disinfectant. This unpleasant smell comes from the chemical trichloroanisole (TCA), which impregnates the wine and is the most common way wine is 'corked'. TCA (and other similar anisole chemicals) is usually formed when the natural fungus in cork reacts with bleaches and chlorine cleaning chemicals used in modern wineries and can infect thousands of bottles. Cork taint can still appear in screw top and rubber corks, often due to other unwanted chemicals in the process. Around 2 per cent of wine is still spoiled by cork taint, although this is getting rarer. But some lucky people find corked wine delicious.

I usually notice there is something odd about a corked wine, though I might sometimes find it still passably drinkable. My wife and daughter can smell the odour at the other end of the table and won't touch it, others don't notice any difference at all. This is mostly a genetic trait: Japanese scientists found that instead of the TCA chemical stimulating unpleasant activity in the aroma receptors, the compound knocked out their electrical signals and induced a strong but fake phantom smell in susceptible people. For the rest of us, the wine is perfectly good and safe to drink. Taste is an individual experience due to our genes and microbes, with no right or wrong answers. If you enjoy a glass of corked wine, carry on!

## Picking the right wine

'Mmm . . . a little citrus . . . maybe some strawberry, passion fruit, and, oh, there's just like the faintest soupçon of like asparagus and just a flutter of a, like a, nutty Edam cheese.' So said Miles Raymond, a character in the 2004 film *Sideways*, but is it rubbish? If our responses to smells and aromas are so individual, then how good are professional tasters in assessing the quality and value of our wine? Most experts can, by looking at the wine and smelling it in the glass, give you a reasonable idea of its origin. But a Bordeaux researcher Frédéric Brochet first raised doubts about the ability to rank wine in 2001. He looked at 143,000 wine-tasting notes from famous groups of professional tasters including Hachette's and Parker wine guides and found the only consistencies came from the colour of the wines; everything else had little agreement on any of the terms, aromas or idioms. Once you removed the effect of colour, by using sneaky dyes, the tasters were lost. He also put fancy and cheap labels on an identical red Bordeaux, and 79 per cent of the experts preferred the posh label.

## Organic, biodynamic and natural wines

Many headache sufferers seek out organic wines to reduce or eliminate sulphites, but they are often disappointed. The organic label is confusing, as it generally means no synthetic molecules or treatments that enter the plant or the roots, but they can use surface chemical sprays like copper sulphate as a fungicide. This does allow the grapes to be organically grown without pesticides, but the method can be traditional and sulphites are nearly always present. Organic wine means different things in different countries: UK organic wine is defined as using organic grapes, with the rest of the wine-making process and sulphite addition unregulated. In France, organic wines are allowed to contain sulphites but at lower amounts. In the US, organic wine means free of added sulphites as well as using organic grapes and yeasts, and no use of pesticides or herbicides whatsoever, similar to Canadian organic wines. In general, organic wines naturally contain up to forty parts per million (ppm) of sulphites, which come from purely natural by-products of the yeast, and normal wines with added sulphites up to 200ppm. It is not just sulphites; increasing use of pesticides in wine-making is causing ecological and financial problems. For example, grape production in France occupies less than 3 per cent of the total area devoted to agriculture and consumes nearly 20 per cent of total pesticides. Organic wines consistently do better in scientific studies for soil quality, grape quality, sustainability and environmental impact, but not consistently better for taste. The last estimate suggested 6.4 per cent of vines globally are producing organic wines. Many of the top wines have converted to these methods, though may not advertise it. Check the small print at the back of the label for details. If you want to support pesticide free vineyards look for a label that says 'wine made from organic grapes' in Europe or 'organic wine' in the US.

### *Biodynamic wine*

A biodynamic wine is part of a holistic vineyard system that is almost always organic with approved traditional leaf sprays, but most organic wines are not biodynamic. There are now over 1,000 vineyards certified by the official body Demeter and numbers are growing rapidly. The biodynamic bar is much higher than for organic, and was an idea started by the philosopher Rudolf Steiner in the 1920s, who with foresight proposed that fermentation could enhance the soil. Biodynamics also employs ancient lunar and planetary calendars to determine the best times for picking, weeding etc., as well as when to rest. According to biodynamic rules, you should never bottle wine or open champagne without precautions on a full moon as too much of the fizz spills out. (You have been warned.) But there's no doubt that the enthusiasts who take it up are good farmers who pay great attention to their soil, plants, weeds, insects and microbes, using standard practices like cover crops and rotation systems. One interesting fact I recently discovered was that each year farmers in vineyards put a special mix of cow dung into cattle horns (commercially known as Preparation 500) and then bury them in a large hole. The idea is to add extra minerals and microbes to the soil; after a while it is dug up and mixed in vortexes in large tanks with 37°C water to increase fermentation. They then spread this ferment across the soil under the vines twice a year, and apparently the results are impressive, both in the quality of the soil and the vines. Although a meta-analysis showed soil quality was better, a recent Swiss experiment couldn't detect differences between the quality of the biodynamic and organic wines over five years. But this trend for sustainable wines looks set to continue as many producers are trying to make the whole process 'organic', using natural yeasts and preservatives and moving away from the 'artificial' manipulation of flavours as well as reducing agrochemicals.

### *Natural wine*

This is an expanding concept that has newly established 'official' French certification (organic plus no added sulphur) and depends on spontaneous fermentation with native yeasts, to maximise the effect of the terroir and minimise chemicals and processing. As biodynamic wines change away from old soil practices, natural wine is moving away from past practices of adding acids, sugar, tannin, chemicals and even colourants to wines to change the balance and look. I tasted some of the fantastic 'original' natural wines on my visit to Georgia, where the grapes are crushed and left to ferment in giant clay amphorae (*qvevri*) in the ground for several months, using natural yeasts, in a method that has been practised unchanged for millennia.

Natural wines are quite different to conventional or organic wines, especially the amber coloured or cloudy unfiltered white wines that have unique aromas and can be very tannic or tart. As the fermentation process with natural yeast is much slower, the flavour profile is similar to that of kombuchas. The lack of sulphur in natural wines can sometimes cause some microbes to proliferate and cause mousy flavours that only a third of the population can detect. I am not fully converted, but like most ferments, it can take time to get used to them and realise they could have health benefits. I do, however, already enjoy many orange wines that are not necessarily organic or natural but leave the white grape skins in contact with the ferment for the same time as red wine, giving more tannin and polyphenols. They go very well if you are tasting dark chocolate. If you are trying to pick genuine natural or low sulphite wines, you need to drink them young and quickly to avoid unwanted oxidation or microbes.

## Low alcohol wines

As alcohol becomes less popular, and hotter summers and grape varieties have pushed up average alcohol contents of wine over the last twenty years to over 15 per cent for many reds, the demand for lower or low alcohol wines has surged. The market is increasing at around 25 per cent a year. Wines made in northern climes, like England and Germany, generally have lower alcohol contents of 10–12 per cent, as do many white wines and sparkling wines. Although yeasts are being engineered to efficiently produce lower alcohol wines, the best-tasting wines I have found are those that are fermented normally and then the alcohol is subsequently filtered or centrifuged out. The latest centrifuge device is called a spinning cone column; it works in a vacuum without overheating the wine and can remove some or most of the alcohol while retaining the aromas and yeast residues. The results provide wines similar in strength to beers at 4–10 per cent (called mid-range) or below 1.2 per cent, classifying them as low alcohol, or if below 0.5 per cent alcohol free. White wines and rosé are easier to play with and not lose the flavours. The New World wines seem to be embracing this trend most enthusiastically, with Australia and New Zealand producing some great-tasting mid-alcohol wines, while I tasted some lovely 1 per cent pinot noir wines from Western Canada that were balanced and not sweet; and there are also good South African and British varieties. But don't be stingy if you are hoping to get a good bottle; the really cheap ones tend to be sugary and unpleasant grape juice. These wines are definitely improving year on year and are part of the future.

Sweet wines are made using residual sugar that doesn't turn to alcohol or by stopping further fermentation and adding alcohol.

There are multiple diverse ways of doing this. They include adding grape juice and stopping further fermentation by filtering or adding sulphur, freezing the grapes before picking (as in German Eiswein) or deliberately allowing a grey fungus called 'noble rot' or *Botrytis cinerea* to grow briefly on the grape. This concentrates the grape and sugar and was first used in Hungary for the Tokaji wine and copied by Sauternes in France. The *Botrytis* fungus perforates the skin and dries out the grape when it gets hot, fermenting the grape on the vine and converting the sugars and acid to produce glycerol, which gives dessert wine its syrupy feel, as well as producing many other fragrant and honey chemical aromas.

Port wine evolved as a pragmatic reliable method of transporting wine from Portugal to England by sea. The alcohol content was increased by stopping the fermentation early and adding brandy to act as a preservative that then killed off any microbes and prevented others from making vinegar. This allowed the wine to be kept for years, which often improved the flavour, complexity and taste. Madeira is similar, but as well as adding brandy it is also heat-treated to 50°C and kept for a few months before cooling. Sherry types like Oloroso are basically oxidised or aged Spanish white wine produced in casks mixed frequently with air – with blends of different vintages.

## Sake and rice wines

The Chinese were probably the first to ferment rice into an alcoholic wine-like drink called *chiu*. They enjoyed it so much that one theory goes that because it caused multiple societal problems, natural selection created a survival or reproductive advantage if a person developed the mutant genes (ADH and ALDH) that break down alcohol more slowly. These gene variants would make them flush and feel nauseous quickly and so be resistant to drinking to excess and alcoholism. Sake is the better-known Japanese version and was imported around the third century, becoming part of national culture. It is made by fermenting rice in water using the natural fungal mould *koji*, which breaks down the starch and natural yeasts to convert it to alcohol.

The grade of rice is crucial for good sake; ideally the outer coating is polished down to the inner core which is the purest element of starch. The fermenting process for sake is performed at low temperatures of around 15°C for 2–4 weeks and extra rice is added. It is then pressed, filtered, diluted and aged for several weeks before pasteurisation and bottling.

On a recent ski trip to Hokkaido, our Japanese mountain guide Yuji introduced us to some amazingly rich varieties of sake, including darker ones brewed with brown rice. They all taste different and can be drunk hot or cold. Although there are exceptions, the best ones are usually savoured cold to extract all the aromas and flavours, and the cheap ones warm. They range from about 14–20 per cent alcohol – slightly stronger than wine. My favourite was called Namasake, which is drunk young as it is a rare 'live' unpasteurised wine and so is rarely exported. The flowery aromas and fruitiness were amazing, especially when you realise the base is just pure rice starch and that it's the microbes that are producing all the complex flavours. It is unclear if pasteurised sake in modest doses has any proven health benefit in humans as the polyphenol content is modest, although anecdotally a few Japanese mice having radiotherapy seem to have done well on it. Many other cultures still use fermented rice drinks, especially in north-east India and Vietnam with reports of mice being healthier.

## *Wine in five*

1. Red wines have high polyphenols and at modest doses of 1–2 glasses per day can have some heart benefits.
2. Our wine tastes are very individual so don't be swayed by price or fancy labels.
3. Headaches due to wine may be caused by sulphites that are lower in organic wines and higher in white wines.
4. Hangovers can be potentially mitigated by fermented foods and probiotics.
5. Resveratrol is in high levels in some red wines and even a glass can be beneficial while some supplements could be harmful.

## Mead, beer and cider

### How to make basic mead

- Buy some good-quality honey (ideally raw, runny and local).
- Dilute with filtered or bottled room temperature water in a ratio of about 4:1.
- Stir together and add to a large glass jar with any optional flavourings (cinnamon stick, one clove) and with a loose lid or cloth cover to allow air in.
- Stir several times each day for 4–7 days to aerate the yeast.
- Strain and drink now or bottle for 2–6 weeks to increase the alcohol content.

### Mead

Mead is probably the oldest fermented alcoholic drink. There is evidence that diluted fermented mead was drunk around 15,000 years ago by the Khoisan people in South Africa. In Europe, the Picts and Celts around the British Isles were making it by 4000 BC.

Mead is just honey wine and is a great base for adding other fruits and herbs to provide extra flavour and wild yeasts. Although honey naturally contains microbes, the honey we eat is generally sterile, either because it has been pasteurised or because the sugar acts as an antimicrobial. Once you dilute it, the brakes are off and the microbes, yeast and bacteria wake from their slumber and start consuming the sugars (glucose and fructose), producing alcohol and carbon dioxide. The key figure is the water content of the honey, which should stay at 17 per cent or below. Just 1–2 per cent more water and fermentation begins.

Honey has many proposed medicinal properties. Although most are unproven, honey can support wound healing, act as an antiseptic and significantly reduce coughs. It also has antioxidant and anti-inflammatory properties, which are probably due to its high polyphenol content along with alkaloids, glycosides and sugars. In terms of microbes, as a hive gathers nectar and plants, it also collects a rich variety of plant microbes, which become enriched in the resulting honey. The honey also attracts other wild microbes, but honey's acidity, lack of water and hydrogen peroxide mean these microbes don't normally cause problems. DNA studies have shown an incredible 191 different organisms in Italian honey, including viruses, fungi, bacteria, parasites, plants and insects alongside alcohol-forming yeasts.

Despite its long history, mead has come in and out of fashion. The only time I saw it in an English pub was when visiting student friends near Southampton. Their local had kegs of mead that were cheaper and more powerful than beer – perfect for the student crowd. I hadn't tasted it since until Chef Justin Horne recently gave me some of his 20 per cent alcohol brew. It tasted so good that I wanted to make my own and include it in the book. It really is very easy – the hardest part is probably getting good raw honey. Pasteurised honey will still work, but it may take longer to ferment, or you can cheat and add a touch of commercial yeast.

## Beer

While working as a doctor in Belgium in the 1980s, I would often take histories from patients admitted with liver problems. When asked how much alcohol they drank, they usually said 'Nothing Docteur, I never touch whisky or spirits.' For them, beer was a soft drink and the 15–20 glasses of lager they consumed daily were just to be sociable. Some Belgian scientists have always claimed that their traditional beers have benefits on the gut microbiome, thanks to the polyphenols and residual yeasts. Although yeasts are dead by the time we drink them, as we have seen with other ferments, they could still have benefits as postbiotics. So picking a cloudier beer with some sediment may be worthwhile.

The beer I brewed as a teenager was strong and unpleasant with strange floating bits, but it taught me the power of *Saccharomyces* yeast to transform a tin of malty substance into something drinkable. Researching the topic five years ago, I found no evidence that the fermenting process for beer had any residual benefits for the drinker. However, this is changing. In general, the darker the beer, the more polyphenols it contains, which should benefit our microbes. Another chemical we encountered in coffee has been discovered in beer, especially darker beer. Called melanoidin, it has potential gut-friendly properties. This has all renewed my interest in beer.

Beer was the original soft drink and probably evolved as a by-product of bread-making in ancient Mesopotamia. Grains rather than grapes are the starting point and barley is most often used for its taste. The main challenge was breaking down the starch in the tough mature grain to allow yeast to convert it to alcohol. Malting was the answer, in which young barley shoots are put in and out of water for a few days to germinate, encouraging the plants to produce an enzyme that breaks down its starch reserves. The malted barley is then dried before being used. For beer, water is added to barley, then mashed up and boiled. Hops – small cone-like flowers of a resin-filled vine (*Humulus lupulus*) – are often added, giving beer much of its flavour, complex aromas and bitterness.

At this stage, fermentation begins and beer can go in different directions. The traditional ale drunk in England for millennia uses the common beer yeast *Saccharomyces cerevisiae*. Fermented for a few days at around 20°C, it is then filtered, clarified, pasteurised and stored in oak barrels for extra flavour before being served at room temperature. It has plenty of flavours and aromas and usually has no foam head. Lager has become the world's favourite variety. The fermentation process is cooler, longer and more expensive, involving storage cellars and a different yeast – *Saccharomyces pastorianus* – which sinks to the bottom. This process produces more carbon dioxide, a foamy head, slightly more alcohol and a crisp taste, without the bitterness or complex aromas of ales. Lager overtook ale sales in the UK in the 1990s, but sales are dropping as drinkers switch to craft ales, IPAs and ciders. Lager has become synonymous with big business, but luckily we are seeing a global switch away from mass-produced beers, with the introduction of many novel brews and flavours. In London in 2010 there were just fourteen independent breweries; by 2020 this had exploded to 129, as demand surged.

Britain and Belgium are the two main countries still producing top-floating yeast dark beers at scale. Belgian beers often still use artisanal methods and brew in the original premises containing the local yeasts; some are still run by monks who originally fled religious persecution in France in the seventeenth century. Despite its small size, Belgium has around 300 breweries, found in areas with plenty of fruit trees. They produce a huge range, from 1.5 per cent beers that were still served in schools as recently as the 1980s, to some wine-like beers of almost 11 per cent, such as lambic and Trappist, with names like Mort Subite (sudden death).

Lambic is called the mother of all beers. It is a spontaneously fermented Belgian beer (meaning it uses wild yeast from the environment) and probably one of the oldest beer styles still brewed in Europe, with recipes dating from the fourteenth century. It is a refreshing, acidic beer with fruity, barnyard, horsey wheat notes and few residual carbs. It is brewed in three stages, taking around three years. The spontaneous natural fermentation occurs when hot water and barley malt are cooled in shallow vats and populated by wild yeast and other microbes that help acidification. The final maturation phase can last for several years where multiple bacteria and many different yeasts play a key role in the final flavour. Geuze beer is a blend of young and old lambics and tends to be less sour. Most traditional Belgian beers will have some residual yeast and dead bacteria at the bottom of the bottle.

In stark contrast to the Belgians, British drinkers are not keen on a full foam head on their beer (except Guinness and stout drinkers). In the rest of Europe, no one will touch a beer without a foam head as a testament to the freshness of the beer. The foam is full of malt and hop proteins, extra taste and aromas – another reason to drink beer in a glass and not swigged from the can or bottle.

## Beer and health

'Guinness is good for you', was a powerful slogan in the eighties and nineties and the Irish media leapt on US research 'proving it' in 2003. When I was a junior doctor in London we were allowed to prescribe stout beers on the wards, to improve appetite, sleep or recovery post-surgery. The reality of the 2003 study was that the drinkers involved were twelve dogs whose clotting profiles improved; real scientists were not very impressed. For the last fifty years, studies strongly linked beer to colon cancer, but over time the evidence has grown weaker. The latest meta-analysis only showed a 6–12 per cent increase in cancer risk for moderate consumption, which was less than for spirits and could be due to error or bias.

In our own microbe studies, beer had a neutral effect on the gut microbiome, at least in moderate amounts. But a big limitation was that we couldn't separate out the different beer styles, leaving the possibility that some types might have benefits, despite the negative effects of the alcohol. Clearly, polyphenols in beer could be important.

Polyphenols are present in beer, but while levels are quite high in dark ales, reaching half that of red wine, lagers are much lower – more similar to white wine. A Spanish study explored the polyphenol question: participants drank beer, non-alcoholic beer with similar polyphenol levels, or gin each day for four weeks. The research suggested beer with or without alcohol improved inflammatory and vascular biomarkers. In another randomised controlled trial, twenty-two adults drank a bottle of either non-alcoholic or regular lager daily. Both showed improvements in diversity of gut microbes, but there was no real control group. Another small clinical trial from Mexico showed the gut microbiomes of those drinking non-alcoholic beers were marginally healthier than in normal beer drinkers after thirty days.

Malt and hops are the source of the main polyphenols in beer, and although total amounts of polyphenols are not that impressive, there is some evidence that specialised hop polyphenols called xanthohumols may be beneficial. When these compounds are given to mice they reduce skin and colon cancer risk, but without any human data, this is wishful thinking. The only human study I found was from the US, where twenty-seven volunteers were given xanthohumol supplements or placebo for four weeks. Sadly, there were no clear benefits either on weight or blood pressure. But the healthy trend for beers with more polyphenols continues. Belgian beers and many modern craft beers contain extra fruits and spices and contain up to 50 per cent more polyphenols than conventional beers, with the highest levels I could find coming from a spicy mix of turmeric, pepper and hops.

Beer, unlike wine, does contain some fibre (around 0.5–2g per can) from the malted grains and hops, which could be good for you and your microbes. While levels vary as much as tenfold, you would need around fifteen cans of an average beer a day to obtain the recommended 'healthy' levels. Studies, though limited, do show that dark beers, stouts and lambics have the highest levels while weak lagers and light beers have the lowest. We think that some non-alcoholic beers have high levels too, but this depends on the method of reducing the alcohol, with thermal dealcoholisation producing higher and healthier levels than methods that restrict fermentation.

Fibre quality is also important; we know that barley contains beta-glucan fibre, which is a food source for many *Lactobacilli* in our guts, so it may help these probiotics survive longer. Studies show that melanoidins, the chemicals produced when barley is malted, stimulate mouse gut microbes, increasing levels of *Bifidobacterium* and *Akkermansia*, which are both considered beneficial. This suggests that many beers contain rarer forms of fibre that have previously not been recognised. A large Spanish brewer tested adding extra soluble fibre to beer to make it healthier. But it worked too well; their internal tests showed that drinkers liked the taste but felt fuller after each bottle, so drank less. Unsurprisingly, brewers did not see this as a marketing breakthrough and quietly ditched the plan.

The obvious downside to beer, apart from the problems of alcohol, is the energy content. Despite recent shifts towards craft beers, we are still suckers for the idea of a 'light' low-sugar beer – Bud Light still has 53 per cent of the massive US market. These beers were heavily promoted in the 1980s, fooling some drinkers into thinking their beer bellies would disappear if they drank enough. Alcohol is metabolised by our bodies differently to sugars and can't be used instantly as energy, so usually ends up stored as abdominal fat. Alcohol, like sugar, provides energy in the form of calories, and light beers have about 30–40 fewer calories per glass, generally at the expense of having less fibre, polyphenols, body and taste.

Low-alcohol beers tasted dreadful when first introduced in the 1980s. The good news is that taste and choice are improving as rapidly as sales. Some manufacturers use special yeast that converts grains to chemicals other than alcohol or can ferment in cold conditions. This means the yeast works slowly, allowing less alcohol to be produced. Alternatively, it can be made by removing the alcohol after normal fermentation, which preserves more polyphenols. I only recently tasted the first alcohol-free beer I actually liked. It was a Bavarian wheat (Weissbier) 0.5 per cent alcohol beer with sufficient complexity to make it pleasant, without being over-sweet. I have since found a range of other enjoyable varieties, including some craft IPAs and Spanish toasted malt lagers (Tostada) that I drink regularly. They don't have to tell you the method of reducing the alcohol yet, so it's hard to rank them on health benefits, but a few have spiked my blood sugar when I'm wearing a glucose monitor. So if you are planning to drink them daily, you should check the ingredients. It's worth pointing out that any alcoholic drink below 1 per cent alcohol concentration is unlikely to have any detectable physiological effect so these are considered alcohol-free in the same way that kombucha is officially non-alcoholic.

While beer – especially artisan types – has some potential health characteristics, I don't think we can call beer a health drink quite yet. But I am enthused by the thought that the dregs at the bottom of my artisan beer contain dead yeast and microbes that have postbiotic properties. I am also excited that food scientists are developing beers that contain live probiotics such as *Lactobacilli paracasei* or the yeast *Saccharomyces boulardii*, ushering in a potential new era for the perfect 'cleansing' ale.

### *Beer in five*

1. Beer is not a health drink, but some dark or artisan types with second fermentations could have advantages that partly balance out the alcohol.
2. Some beers have high fibre, high polyphenols and other postbiotic fermentation products.
3. Beer drinking has been shown to improve gut health in a few short studies.
4. Low-alcohol beers containing polyphenols appear optimal for health.
5. Low-alcohol beers that use thermal methods to reduce alcohol after fermentation may be healthier.

## Cider and perry

The ancient history of cider is obscure but probably predates the Bible. All it takes is a few ripe apples left in water and a bit of warmth to allow microbial fermentation and alcohol to form. Although the Romans may have picked it up from the Celts, in Europe, the Basques are generally credited with spreading it further from about the sixth century. Each region of Europe had its own apples and distinctive style of cider-making, which are still evident today. The world's biggest producer is now the West of England, which recently outgrew Normandy. Thanks to a worldwide revival, cider is made in most countries where apples are grown.

Cider is best made using a blend of sour, bitter and semi-sweet apple varieties rich in tannins, often with a reddish-brown skin. The process of fermenting is similar to wine-making but simpler and slower. The fruit is crushed in cider presses into a pomace, filtered, put into vats and left at cool temperatures for a minimum of three months, using either natural yeast from the apple skin or added *Saccharomyces* yeast. It is then siphoned off into a fresh vat for a second fermentation with extra sugar to give it some fizz. If bottled, the mix can be blended and more sugar added.

The amount of alcohol is highly variable (up to 12 per cent), often depending on local taxation. Some of the rough, traditionally made cloudy West Country ciders that are the most alcoholic are called scrumpy, derived from the old name for withered apples. These scrumpies used to be served in industrial 3-litre plastic containers adorned with 'poison' stickers. I encountered these when I was a reckless and poor student, and still remember the hangover, but it's good to see they are now considered gourmet products with the name and region being protected (like Champagne). The skill is to blend the right apples with selected yeasts to achieve the perfect balance of acidity, tannins and sweetness, without worrying too much about clarity or colour.

One study looked at six artisanal ciders made with spontaneous fermentation without starter cultures. In addition to the common *Saccharomyces* and *Brettanomyces* found in beer, the yeast diversity was even greater and included over twenty non-conventional species. The bacterial community was also more diverse in cider than beer.

The UK cider business accounts for only 7 per cent of alcohol sales but is the biggest in the world at nearly £3 billion per year in sales. While high-end artisanal brands specifying regions and orchards are booming, most sales are still in cheaper, mass-produced ciders. These are heavily filtered, purified and usually pasteurised and can be nearly colourless or yellow. White cider is an industrialised version that is heavily refined, has little relationship with the original apples and has a high alcohol content of around 8 per cent. Mass-market ciders often contain high levels of sugar compared with drier artisan varieties or beer. A third of the market is now brightly coloured, high-sugar fruit ciders that often don't contain much apple or alcohol but contain around twelve spoons of sugar per pint. This brief trend for even sweeter commercial ciders or fruit alcohols seems to have peaked, thankfully.

Cloudy or unfiltered ciders are generally considered healthier by the consumer, though many prefer the purified versions. In general, the cloudiness signifies small particles of pectin fibre from the apple that haven't been broken down, as well as many yeasts and microbial debris. Filtering or fining the cider makes the flavour more stable but may remove some of the tasty and beneficial chemicals like the tannic polyphenols and other sources of fibre.

The Spanish, French and Belgians have traditionally drunk cider in Champagne-style bottles with seafood, crêpes or foie gras. With alcohol levels of around 8 per cent, it's enjoyed more like wine. Pear cider is made from pear concentrate added to apple cider, whereas a more traditional (and nicer) option is the natural pear juice equivalent called perry. Fashionable ladies in the sixties and seventies drank Babycham, which came in mini champagne bottles and – unknown to me – was actually sparkling alcoholic pear juice. Thanks to clever marketing, most people thought it was cheap champagne. Sales have since declined, but artisanal perry has had a revival in recent years.

## Cider and health

We all know apples and pears are good for you, but what about the fermented forms of cider or perry? There are no good clinical studies in mice or humans, which is strange when you see the vast literature for apple cider vinegar, so we must speculate. The average amount of polyphenols in an equivalent portion of artisanal (not industrial) cider is slightly greater than red wine, and over three times the equivalent glass of lager. As usual, the fermenting process with the microbes produces more chemical polyphenols than non-alcoholic apple juice. The cloudier your cider, the more likely it contains high amounts of polyphenols and the better it may be for you. In the cloudy haze of real cider, a study found that if you remove the cloudiness, you lose polyphenols (mainly procyanidins), which were the main compounds identified and accounted for at least 31 per cent of the haze. Polysaccharide sugars, many of which have some prebiotic fibre potential, accounted for around 10 per cent, with additional potassium and calcium as the main minerals. So the healthier cider is probably the cloudy one.

### *Cider and perry in five*

1. Fermented apple cider has more total polyphenols than red wine.
2. Drinking craft cider in moderation has potential as a health drink.
3. Industrial cider is heavily refined and purified, removing most beneficial chemicals.
4. Perry and cider are easy to make at home if you skip the fruit-pressing stage and use juice.
5. Drink your cider and perry cloudy for maximum health benefits.

## Condiments and fermented spices

### How to make honey-fermented garlic

- Peel enough garlic to fill a jam jar by two-thirds.
- Add good-quality honey (ideally raw) to cover the garlic.
- Close the lid and turn it every few days for 7–14 days.
- Watch it ferment and transform into a clearer liquid.
- Then sieve it and eat whole or blend it as a salad dressing.

Many condiments we love to use today originated as fermented foods. We have already talked about some of the major ones such as sauerkraut, kimchi and soy sauce, but there are many others including Marmite, Tabasco and even tomato ketchup.

Europe's original condiment was probably the fermented fish sauce (garum) used by the Romans, and before soy sauce became popular in China and many other Asian countries, fish-based sauces were in common use everywhere. But vinegar-based sauces have also been around since ancient times, across the world. Fermenting vegetables doesn't get any easier than honey garlic (see above) – garlic naturally comes complete with its own microbiome starter and chemicals such as allicin that deter other microbes, which some cultures believe gives it extra health benefits.

## Tomato ketchup

A bottle of this is one of the commonest sights in Anglo-Saxon cupboards and a permanent resident on many café tables. It is one of the earliest processed foods and has been mentioned at least since the seventeenth century. Ketchup scholars believe it may have originated from a dark fermented fish sauce from China or Malaysia (*ke tsiap*) or even kecap manis from Indonesia. Whatever the precise origin it is likely that English colonists spotted it and brought it back to the UK and to North America. Food historians have found that in 1700s England there were three main fermented sauces that were popular: mushroom, walnut and fish – and they were all called ketchups. Mushroom ketchup was the most popular and recipes at the time describe how mushrooms were salted overnight to lightly ferment them and concentrate flavours and then simmered at a low heat with herbs and spices (e.g. horseradish, cloves, mace, nutmeg and allspice). Other recipes suggest they were salted and then added (with spices) either to a homemade vinegar or to a salty brine, later removing the mushrooms and bottling the runny sauce. The umami flavours were probably similar to the older fish sauces they replaced. Blending together different ketchups was common in cookery books of the 1750s; the original Worcestershire sauce was a blend of walnut and mushroom ketchups. Sadly the modern version is no longer a fermented product, although it still has properties that make it a great addition to many dishes.

The first tomato ketchup recipe appeared in 1812, changing the main ingredient from mushroom to pickled tomato and slowly evolving it into its modern form. In the US, many were eating ketchup well before fresh tomatoes became cheap and popular. The first mass-produced product was made in 1837 followed by Henry John Heinz's version forty years later that in its modified form is still the market leader in most of the world, with industrial vinegar replacing the fermented versions of the past. In the UK, Heinz, like others, has a 'clean label' of contents in order of concentration; tomatoes, vinegar, sugar, salt, herb extracts, spice and celery (to provide the nitrite preservatives). US versions have high-fructose corn syrup, as well as other chemicals as their health agencies are less worried about 'natural' flavourings, but all bottles have some chemicals to help the flow. A single tablespoon portion of modern ketchup is equivalent to a teaspoon of pure sugar (4g). Low sugar versions are available, but at the cost of many more added chemicals and processing.

You can now recreate the old-style fermented tomato ketchup easily with some tomato purée, soy sauce, salt and herbs to your own taste (see page 241).

## Tabasco and fermented peppers

Chillies and peppers belong to the *Capiscum* family that have plenty of healthy polyphenol chemicals, which may be the reason for studies that show it reduces inflammation and blood pressure. They also ferment well, breaking down the raw intensity into more subtle and complex flavours and extra polyphenols. They naturally carry plenty of microbes to kick-start wild fermentation such as *L. plantarum* and *Leuconostoc*. One of the best-known fermented chilli products is Tabasco. The history of Tabasco red pepper sauce is still hotly disputed, either coming from Louisiana in 1868 from the present company's founder Ed McIlhenny, or some nineteen years earlier from a rich plantation owner in New Orleans who planted special Mexican tabasco chillies on his farm and prepared his own sauce. Peppers are ground into a mash and placed along with salt in white oak old whisky barrels to ferment the mash for up to three years, then strained, diluted with distilled vinegar and bottled as a finished sauce. The original Tabasco pepper sauce still contains peppers fermented for three years.

Although you are unlikely to eat enough chilli paste or sauce to reap any health benefits, one small study has shown that low to moderate fiery chilli pastes (but not hotter ones) can have some effects on reducing blood lipids and inflammation. I was surprised how easy it is to make your own mash to use as a condiment and adjust to your own levels of spiciness. I mixed up some red peppers, chillies and garlic from the fridge in under ten minutes to make my pepper and garlic sauce.

I love this mix and find I now make it regularly and either have it as a relish with cheese or add it to countless other recipes such as pasta sauces, to give them extra depth. You can roast the chillies or peppers beforehand if you want even more flavour, and adding green peppers also works really well.

## Other vegetable pickles

Some cultures (especially Ukraine and Romania) ferment tomatoes, onions and other plants whole – this protects their structure but requires longer fermentation times. In Turkey, they like to add some vinegar as well as brine to make turşu. Do try fermenting different krauts or kimchi variants yourself or look up precise recipes for ones you fancy such as beetroot borscht, *cortido* (onion, carrots, chillies, lime and oregano), red cabbage and apple kraut, or if you suddenly have too much kale or seasonal wild garlic, just ferment them with salt so they don't go to waste.

### *Fermented mushrooms*

It was hard to find many modern recipes for fermenting mushrooms, which are one of my favourite healthy vegetables/fungi. This is because they can lose structure and often end up with an odd or unpleasant taste. Neil Rankin, the ex-barbecue chef behind Symplicity Foods, gave me a tip that chestnut mushrooms work well if quickly fried first to produce a Maillard reaction on the outside. The resulting salt fermentation gives a much more balanced umami flavour and can be mixed with onions and miso. After some trial and error I found that dry-frying some round or chestnut mushrooms to brown them and then fermenting them in salty 2 per cent brine with garlic worked a treat.

### *Pickled gherkins*

I loved these as a kid, eating them from a massive jar that my Jewish grandmother kept in the cool larder. I don't remember if they were made in brine (and so with microbes that were still alive) or in vinegar as most of them are now. While living in Belgium I got to love the tiny baby cornichons that go so perfectly with ham, salami, cheese and mustard. Cucumber and its close cousin the gherkin come from the same *Cucumis sativus* family, but for pickling you need a combination of a tough skin without too much bitterness. In the UK and US the thin-skinned Dutch cucumber predominates our stores, but in southern Europe you find the thicker-skinned variety. Having recently pickled an English cucumber to a horrid mush, I tried pickling an organic, thick-skinned Spanish one. I quartered it and sliced it into long chunks and added 2 per cent salt, peppercorns, coriander seeds and dill. To reduce the enzymes breaking down the fruit too quickly, I added some antioxidant in the form of a regular teabag (bay leaves also work) and topped up the jar with more salty water. The result after three days was much better with a decent crunch. If you do find some large pickling gherkins that fit in a jar, remember to pierce the centre with some needles so the microbes in the brine can penetrate and the fermentation is even. As they continue to slowly ferment in the fridge, it is best to eat them within a week as mine got slowly soggier.

### *Fermenting legumes and spices*

Beans, peas, chickpeas and lentils are easy to ferment and already very healthy to start with but need to be fully cooked before fermenting with 2 per cent salt; a process that takes about three days. The extra fermenting unlocks even more nutrients than in the cooked versions. As the natural LAB will be killed off you will need to add a starter of some kind, which could be a previous salty brine used for fermenting or a spoon of milk or water kefir. You can make fermented hummus this way, but add the tahini and olive oil afterwards, as fermenting any kind of oil is tricky and best avoided.

Spices are a great source of diverse polyphenols and should be part of any healthy diet. One I use a lot in cooking is turmeric (*Curcuma longa*); it has long been used in Chinese culture and Ayurvedic medicine as an anti-inflammatory medication. Curcumin, the main component, is an antioxidant and anti-inflammatory compound; in multiple, small and as yet unimpressive studies it has been suggested that curcumin can help with nearly every common disease, from osteoarthritis, type 2 diabetes and high blood lipids to dementia and cancer. Its potential health benefits make it interesting, but what if you could further boost these properties by fermentation? Microbial fermentation has been shown to significantly improve the availability and curcumin levels in turmeric and boost its pharmacological effects. Studies have shown improvements on blood levels of blood fats and inflammation when fermented with LAB bacteria, camel milk and natto. So, as the taste of fermented turmeric is not very different to the mild taste of the original, you should add it to your repertoire (even if you can't get hold of camel milk). I now add some chopped or powdered turmeric to my kombucha and water kefirs at the secondary fermentation stage. But turmeric does ferment easily by itself in a simple sugar and water solution after a few days (see page 249).

There is little research on other spices that have already been shown in dry form to be beneficial for the gut microbiome, but there is no reason to believe that fermentation would not also increase the bioavailability of their healthy components. We should be doing more research and I have started adding more spices to my ferments while they are at room temperature.

## Marmite and Vegemite

The Marmite effect has become a marketing term to describe either loving or hating something. We explored this over twenty years ago in a twin study and found much greater similarity of Marmite preferences among identical twins, showing it had a large genetic effect, which wasn't surprising. But as usual, genes didn't work alone; there was also a strong effect of environment and early exposure to this very original, strong, salty umami flavour of concentrated brewer's yeast. Both Marmite and the Australian version Vegemite are produced as by-products of the beer industry, the only difference being that Vegemite is thicker and has a few added vegetables and spices, plus some additives. Naturally, Australians are convinced theirs is better.

They are both produced by the fermenting process of beer by yeast, but studies have shown that after the heating stage both were completely sterile products at point of purchase. This means they are not probiotics or of any use as starters unless you add in some more brewer's yeast. But they still contain many healthy proteins and vitamins, including high levels of B12. This means a portion can provide 25 per cent of the daily allowance, making it useful for non-meat eaters. These ferments should not be confused with products such as Bovril, which is primarily a salty beef extract with only small amounts of yeast.

Researchers have fed their worms most things over the years, but Marmite was one of their diet ingredients that increased longevity. Another ambitious theory was that Marmite could have an effect on human brain activity and mood via some of the chemicals produced by the yeast. One of these is GABA, which has a relaxing effect on the brain and is increased in the early stages of drinking alcohol and taking a Valium (benzodiazepine) tablet. Yeast-produced GABA could be more effective than consuming GABA supplements as these cannot easily pass into the brain and are likely to be broken down in the digestive system. Researchers tested twenty-eight healthy volunteers, half of which were assigned to the active Marmite group while the rest were peanut butter controls. They measured electrical activity in the brain and found eating a teaspoon of Marmite every day for a month had a calming effect. The subjects also showed a 30 per cent decrease in their brain's response to visual patterns. This is possibly the first dietary product to alter GABA levels, so if the study is replicated it would be very exciting. So Marmite or Vegemite brewer's yeasts could have some health benefits, including calming you down after a tough day at work – and this despite the brewer's yeast being completely dead.

### *Condiments and fermented spices in five*

1. Fermenting a spice increases its polyphenol content and chemical properties.
2. Fermenting is an excellent way of preserving long-term condiments that last months.
3. Homemade tomato ketchup is healthy and simple to make.
4. Try fermenting your own spices such as turmeric to increase its health benefits.
5. Marmite or Vegemite contain large amounts of brewer's yeast, which could have calming effects on the brain.

# PART FOUR
# What Next?

## The future of fermentation

I ordered my veggie pizza at Homeslice Pizza, a small, fashionable establishment in Marylebone, and bit on the delicious spicy nduja (a spicy Italian meat) topping, after being assured it was vegan and made of fermented vegetables. My meat-loving companions were equally fooled. I was astonished that you could ferment vegetables to replicate meat products so well. Of course, I had to find the producer, Symplicity Foods, in north-west London. When I saw their vats of fermenting tomatoes, soy, onions and mushrooms, and the patties that came out the other end to make sausages, burgers, mince and ragu, I was astonished by the simplicity of the process and how little interference the product needed to taste so good.

They've now developed over ten different products and are supplying restaurants with a range of foods, like their meat-free mince made with a base of fermented chestnut mushrooms, beetroot, onions, and miso sauce seasoned with fennel seeds, garlic, red chilli flakes and olive oil. I assumed there would be plenty of extra additives and flavourings to give it that umami meat taste, and emulsifiers and binding agents to add texture, but I was wrong. The fermenting of the microbes had done all the hard work and this was relatively unprocessed. The end product was pasteurised – so sterile – but had plenty of fibre, polyphenols and other postbiotic products. Importantly, it tasted great without the need of any emulsifiers, sweeteners, thickeners or artificial additives that are found in most meat alternative UPFs. This is what fermented meat alternative foods will look like in the future.

The demand for all fermented food and drink products is rapidly growing – the market will be worth nearly $100 billion by 2032 and is currently growing at 7 per cent a year. A new field called precision fermentation will contribute much of this, predicted to grow at

48 per cent a year. (It is actually not that new; it has been a discreet industry, keeping quiet about its methods to avoid upsetting the public, who might regard it the same way they did GM foods.) Interest in precision fermentation has grown with discovery of the gut microbiome and genetic engineering (GE), and with big advances in fermentation technology. But the reality is that we all already consume products of precision fermentation without realising.

## Precision fermentation

The first example of precision fermentation was by a genetics startup, Genentech, using *E. coli* bacteria to produce human insulin back in 1982, which was a huge breakthrough. I remember as a junior doctor dealing with diabetes patients who had developed antibodies to traditional insulin, previously extracted from pigs, and this novel fermented product – identical to human insulin but made by microbes – changed their lives. Since then, the industry has used fermentation by genetically modified organisms to produce many diverse products that are components of foods we all eat today, from cheese to tomato sauce. Most people also have no clue that the majority of vitamins and supplements on sale today are made not by some organic workshop or extracted from fresh hand-picked fruits but from microbes in giant vats using precision fermentation.

Engineered microbes now routinely produce substances that originally came from animals, including 90 per cent of the rennet used worldwide in cheesemaking, reducing the need to slice open some poor young animal's stomach, making cheeses such as Parmesan less of an ethical conundrum for vegetarians. Why any cheese manufacturers (the remaining 10 per cent) still use animal rennet when we have a bio-identical version that isn't cruel or bad for the planet seems baffling. Other common examples of products made by precision fermentation include vitamin B2, and vitamin C or citric acid used as a food preservative, and there are many others such as resveratrol (an antioxidant supplement, see page 164), stevia (a sweetener), psyllium seed husk (to reduce cholesterol), soluble fibres, sulforaphane (an antioxidant from broccoli) or isoflavones from soy. Fermentation can also produce valuable pharmaceuticals, such as the anti-diabetes drug sitagliptin from Merck, or in agrotech the nitrogen-fixing fertiliser for corn called Proven by Pivot Bio.

The US brand Impossible Foods already uses an engineered yeast *Pichia pastoris* to produce a protein called heme to give its plant-based burger the flavour and colour of animal meat. The heme normally comes from a fungus found on the roots of soy plants. The Impossible burgers and meatballs have been popular with vegan and meat-eating consumers, but the burgers include many of the hallmark ingredients of ultra-processed food. But these should be seen just as prototypes for the future. While the present products are merely copies made more ethically and sustainably they are not necessarily healthier. But as precision fermentation develops, we should in the future be able to recreate meat taste and texture by combining precise manufactured animal proteins and fats in a healthier way, without resorting to cheap chemical additives, emulsifiers, gums and flavourings.

## Fermented proteins

There are several factors at the heart of improving both existing fermentation and the new area of precision fermentation: selecting the best strain, understanding the ideal metabolic pathways needed for growth of the microbe, and assembling the genes needed and placing them into a microbial framework. The technology relies heavily on our increasing understanding of the microbial genomes and their metabolic functions, and advances in genetics and big computing mean that the functions of specific strains of microbes can be predicted, so that the optimum mix of strains can produce, say, lactic acid, alcohol, or vitamins much more efficiently.

Replacing just 20 per cent of animal protein with fermented proteins would halve our agricultural carbon emissions. Non-animal precision fermented milk has 97 per cent fewer greenhouse gas emissions, and 99 per cent less water usage than dairy milk, can avoid the lactose that many are intolerant of, and is now already commercially available in some countries. Although the price in 2024 was nearly

$7 per litre, it is just a question of time until the price reduces below that of the animal milk products. In 2023, there were more than eighty-nine startups using precision fermentation to create many different products, from dairy proteins to chicken-free eggs, food dyes, vitamins, and fungi-sourced fats to bring juiciness to plant-based burgers. Others are scaling up fast using vast solar-powered bioreactors in Oman and Texas to grow engineered mushroom mycelium that ferments pea and soy protein into high-protein meat substitutes. The ones I tasted at the tech fair were not amazing, but in simple ready meals, burgers or meat sauces they would be fine. They hope to reach price parity with beef patties very soon and already have EU approval as a novel food.

Sales of milk protein whey are booming, so bodybuilders can get their fix without having to milk 270 million very inefficient dairy cows daily. Powdered whey protein is now made with beta-lactoglobulin using undisclosed fungi and although technically a UPF, it will still be better for the planet than consuming dairy. Egg whites are big business in food manufacturing and simple to precision ferment. As the price of eggs has risen and supply is threatened by bird flu and other infections, egg whites are the first fermented protein to cost the same as the animal version – and they are biologically identical. Why would you want to keep billions of laying chickens in cages polluting our fields and rivers if you can have the identical product cheaper? Most eggs get used as minor ingredients in food production to add protein and bind foods together. Egg yolks have so far proved trickier to produce and will take a few more years to crack.

Early adopters approving novel products include the US, Israel, Hong Kong, the Gulf states and Singapore and others are following. The UK and Europe are lagging behind, perhaps too susceptible to the pressures of farming and the traditional meat and dairy lobby. Another reason may be the continuing reluctance of the public to endorse genetically engineered foods, although no scientist I know has any real concerns any more. Technically, precision fermentation is not a GMO or even genetically engineered, as the proteins themselves are unmodified, and only the producer (i.e. the microbe) is being modified but never eaten in the final product. This is quite different to the original genetically modified tomato. If the public can start to see healthy products at price parity with meat and dairy, just small uptakes could have considerable impact on climate change.

The technology is also advancing for infant formula, which currently relies on expensive, climate-damaging animal protein and does nothing to nurture the infant microbiome. Helaina, founded in 2019, genetically engineers yeast to produce proteins found in human breast milk. These proteins, and the tiny complex sugars attached to them (also known as human milk oligosaccharides, or HMO) could aid immune-system development, partly by feeding the beneficial bacteria in the infant gut and avoiding some of the problems of cow's milk. The products are made in enormous vats, purified and then spray-dried to form a powder. Companies that use cultivated cell methods face a tough road to get approvals, as quite rightly novel infant feeds need to show they are safe and provide complete nutrition. Combining these with breastmilk microbiome probiotics by companies (such as UK startup BoobyBiome) could bring us closer to a true human breastmilk alternative. Other companies are fermenting key components of human milk, such as microbial origin HMO sugars for nutrition and lactoferrin, which provides immune defences against infection and iron regulation. We will soon see these components of human milk appearing in infant snacks, but it will be a very long while before we can replicate the complexity and benefits of human milk (if ever). Compared to current ultra-processed, HMO-free, animal infant formula alternatives that do nothing to nurture the infant microbiome, this is a massive step in the right direction.

The whole field of creating alternative proteins (including plant-based methods, cultivated meats and fish) and precision nutrition has attracted over $14 billion in funds by 2022. Nearly $3 billion was invested in precision fermentation in 2021 and 2022, with investment growing at 19 per cent a year over the last decade. As the technology and market increases rapidly, it is hard to keep up, and surveys show that consumers are confused by the products and labels. There is now a trend towards highlighting the fermentation process and no longer calling them 'synbio' (synthetic biology) or 'animal-free' products but instead using the friendlier 'fermented milks' or 'dairy or whey protein isolates produced by fermentation'. Surveys of the public in the UK and Canada have shown a willingness to try milk alternative products more readily than meat alternatives.

## Fermented fats

At a food tech fair in London I tasted some good-looking vegan pork belly made from plant proteins mixed with strips of crispy fat by a San Francisco company called Lypid that uses a secret method. It tasted pretty reasonable – much better than current vegan offerings – and you can buy it online. Currently, recreating the taste of fatty meat requires manufacturers to use palm oils or coconut oil with plenty of artificial flavourings and because the melting point of plant oils is lower than animal oils, they are often greasy, smell of coconut and need binding agents to hold it together.

To address this problem, a few companies have now managed to get yeast to produce animal fats: one from Sweden is focusing on beef fat, because of the huge climate impact. They already have a prototype meatball and are scaling up. Hoxton Farms in London are producing beef and pork fat, but rather than using genetically modified yeast, they use cultured real animal cells to produce large amounts of pale animal fat. In time, fed on plant-derived proteins, nutrients and growth factors, they multiply to form a mass of fat cells. The fat can be grown in healthy environments, encouraging uptake of omega-3 fatty acids and potentially other beneficial chemicals. Cubiq Foods is another cell-based company in Barcelona now backed by US giant Cargill Inc, who are looking to mix these fats with plant products to produce the near perfect vegan burger. Both the cell-based methods and the yeast-based fermentation methods use equipment similar to industrial beer makers so the likely future is that these two technologies will merge.

Cultured and precision-fermented fats are much simpler than proteins because fat is a simple blob-like cell that enlarges when storing energy and sticks together in a bioreactor – just like it does inside our bodies. Early tastings of fake bacon suggest a mix of just 10 per cent animal fat to 90 per cent plant protein can be enough to fool our palates. As other animal fats and oils are produced this way we will see increasing mixes of plants and alternate proteins. Talking to some of these companies I heard a consistent message. They want to produce foods that are healthy for humans and the planet and avoid the mistakes of the big companies making highly processed foods with multiple unwanted chemicals, even if this means taking a bit longer to get there. Let's hope this happens.

By pushing the boundaries of precision fermentation we can create many exciting possibilities for health and nutrition. The many start-ups are secretive, but we do know of some examples such as red yeast *Rhodosporidium toruloides*, which has been engineered to produce a range of fats, polyphenols and pigments, as well as novel compounds that make nutrients more accessible. This yeast alone has over 150 specific genes that can be manipulated to produce fats for our foods or medicine and there are many thousands of other species. This means we can start to engineer novel proteins that could be useful, such as anti-ageing proteins or others that mimic gut hormones and so act like the blockbuster anti-obesity GLP-1 drugs like Wegovy or Mounjaro.

Many startups are just trying to get microbes to produce high-quality protein in the most efficient, sustainable way possible. Some bacteria naturally contain high levels of protein in their cell walls that could also be used as food (in the same way other companies are growing crickets or larvae). These natural protein powders could be used in all processed foods, including meat substitutes, yogurts, smoothies, pet food and – importantly – in animal feed, but in a vastly more sustainable and efficient way. This is precision fermentation but using detailed knowledge of the microbes and tweaking their environment rather than their genes. Microbes are being discovered that could live off alternative sources of sustainable energy. One of these is a hydrogen-loving bacteria (probably *Xanthobacter* from the Baltic ocean). These bacterial cells, when harvested from giant bioreactors and dried, consist of around 65 per cent protein, 5 per cent fat, 25 per cent carbs and fibre and a few minerals including potassium. This is very similar to soy protein, but at least a tenth of the environmental cost and a minute fraction of the impact of beef. Another company farms an ancient bacteria called *Cyanobacteria* and claims to have made over eight meat proteins that can provide the structure, taste and texture of meats. These methods could be game-changing if scaling of these bioreactors works and other willing microbes can be found that thrive on substances that are essentially free like sun, air and water.

## Fermented plastic

As well as hormones, vitamins, novel medicines, protein, fats and antioxidants, fermentation could even solve issues of food and packaging waste. Fermentation is now being used to break down plastics as well as make new, biodegradable ones. In Australia, scientists are feeding bugs seaweed to make biodegradable plastics, while a group of scientists from Cambridge university showed that the bacteria *Ideonella sakaiensis* can break down a PET plastic bottle in six weeks and even make electricity in the process. The combination of fermentation potentially reducing pollution from livestock, plus breaking down plastic pollution to create energy, plus making a biodegradable plastic from seaweed makes for a very positive planetary equation.

## Novel probiotics

While the above are great ways for us to get non-animal protein to consume, let's not forget about producing and discovering more probiotics, postbiotics and prebiotics. As traditional dairy gets more expensive for individuals and crippling for the environment we will be seeing more imaginative creations at home and in the shops. We already have non-dairy kefirs, made from coconut or soy milks, but not everyone likes the taste. I love the idea that kefir made from microbe-produced whey protein milks, maybe with cultured animal fats, will become commonplace. I should soon be able to eat some of my favourite cheeses without the guilt of their carbon footprint.

I think we will also see more synbiotic foods, by which I mean combining prebiotics (fertiliser) and probiotics (seeds) together – and maybe postbiotics too. This is the essence of traditional fermented foods like sauerkraut and kimchi, but using modern technology we can add probiotics and postbiotics to all kinds of novel food substances to improve their lifespan and chances of colonising our guts. New techniques can now modify probiotic fermenting bugs like *Lactobacillus* to detect infections like cholera and both destroy the cholera and alert the patient and doctor to the infection, acting as perfect prevention. Other labs are working on modifying bacteria to produce cancer-killing probiotics that seek and kill early cancer cells. You could find microbes that are attracted to cancer cells and then modify them to produce ammonia, which in turn then attracts our immune T cells to destroy them and eliminate the cancer.

The next wave of probiotic microbes are coming from discoveries within our human guts, rather than, as in the past, from tough microbes in traditional fermented foods that are not sensitive to oxygen. This means that we have to try and find a way to keep them alive without oxygen, until they can reach the colon. If this can be done they could be given as capsules or added to foods. In 2023, a team did this by finding out that two different gut microbes tended to be best buddies. One was *Faecalibacterium prauznitzii* (*F. prau*), a healthy bug that eats fibre, produces butyrate and reduces inflammation, and the other was *Desulfovibrio piger*, which didn't eat fibre but enjoyed lactate and produced acetate that *F. prau* could feed off. They survived better together than alone. The team then slowly trained *F. prau* to be less sensitive to oxygen; after about ten generations, a strain had evolved that could survive long enough to be scaled up in large amounts and put into humans for a trial (alongside its buddy *D. piger*). The early trial in humans was a success and showed that it could potentially help people, especially those with low levels of *F. prau* at the onset, thereby offering a personalised approach. Another novel bug is called *Anaerobutyricum soehngenii* that lives off small fatty acids rather than sugar and was found in a human pilot study to reduce blood sugar levels.

With our huge and growing ZOE database of hundreds of thousands of volunteer microbiomes, we have likely many blockbuster microbes waiting to be discovered and cultivated. Two of the three top microbes we found recently to be most strongly linked to health and food were totally unknown bugs with enormous potential. All these inventions could one day be part of our food and everyday lives. Our recent understanding of the power of modifying microbes is probably akin to the excitement felt in the Fertile Crescent when they first discovered what you could do with fermented milks.

The future of the traditional fermented field is not without challenges. There are still fears of microbial infections and a resistance of regulators to officially accept health claims. Of the over 400 health claim applications made for probiotic and fermented foods, only one has been authorised so far by the European Commission. That lonely claim is a pretty pathetic one: it is that 'living cultures in yogurt improve lactose digestion in those with lactose intolerance'. All other gut functions and immunity claims have been rejected because of apparently insufficient scientific evidence. Since Brexit, instead of going independent and making it easier to progress probiotics science, the UK government has made it harder. Unlike the laxity for health claims for vitamin supplements, they require the same level of evidence as for pharma trials for probiotics, which usually cost over a billion dollars each, which no food producer can afford. So instead of making claims about the probiotics themselves (and funding research), the industry is forced to add minute amounts of additives like zinc, manganese or calcium to products so they can legally claim immune or gut benefits based on outdated and spurious claims that no scientist believes. The US and most other countries are only slightly better. The regulation and promotion of gut-friendly foods is a mess.

In 2024, I helped the ZOE team design and launch a gut health shot with a UK supermarket chain, designed to be the best on the market. Using knowledge of the microbiome we combined a multi-strain commercial kefir with fourteen bacterial strains (we confirmed this by sequencing) plus multiple fruits and fibres to provide both high fibre and polyphenol levels combined with beneficial strains, as well as low sugar and no harmful additives. For the first few weeks it was outselling water. We are likely to see many other similar combinations of microbes and plants in future. Matching probiotic microbes with the best food matrix to increase survival is a big emerging field, especially as we move away from dairy to plants. Sometimes the plants don't provide as much protection against long-term storage and gastric juices, but this can now be achieved by microencapsulation technologies, in which microbial cells are entrapped into food with a protective coating. Recent examples include encapsulating the microbe *L. reuteri* into a probiotic tomato juice and *L. plantarum* into dark chocolate. We mustn't forget the use of dead microbes as a health food either; the postbiotic field is in its infancy and already there are several tech startups focusing entirely on developing novel foods based on this concept. We plan to update our ZOE Daily30+ wholefood prebiotic supplement by adding some postbiotic kombucha. As more studies confirm the benefits of dead microbes, this could be a huge new area of development as there are fewer safety issues as it is a food product not a live organism.

## Personalising probiotics

The other big future development will be personalised pre- and probiotics. As we all have unique gut microbiomes and even at species level only share around 11 per cent of our microbes with strangers (in contrast to DNA where we share 99 per cent), it makes sense that we should not all be blindly given the same medicine. A few studies have shown that the response to probiotics depends on the components of your own microbiome when you start treatment. Some companies (such as Sun Genomics or GUTolution) are already offering this service, although they are still using crude measurements for selection. I tried taking the novel probiotic *Akkermansia* available in the US for two weeks and tested my microbes with more advanced metagenomics at ZOE. I had very low levels of *Akkermansia* at baseline, but, reassuringly, levels increased sharply two weeks later and declined when I stopped, which is exactly what I expected and why precision probiotics are a big market.

Personalisation would mean that you wouldn't need to take

expensive supplements of microbes that you already have plenty of, nor take prebiotic supplements that feed the wrong microbes. Instead you could take precise amounts without guesswork or waste. How quickly this trend for personalisation will extend to the general consumer without a gut test is unclear, but with the advances in precision fermentation we could soon have special yogurts for people with depression, kimchis for high blood pressure, or a cheese to help with sleep or menopause problems, all made without harming an animal. It's great to see cutting-edge science combining with traditional fermenting techniques to provide novel medicines and foods.

### *Future of fermentation in five*

1. The use of postbiotics (dead microbes) and personalised pre- and probiotics as medicinal foods will grow.
2. Fermented plants will play a bigger role in replacing unhealthy meat products as prices come down.
3. Precision fermentation by microbes already produces insulin, rennet in cheese, replica egg whites, milk powder and whey protein sustainably.
4. In the future, precision fermenting can produce key proteins, novel anti-cancer medicines, novel foods and help the planet.
5. Microbial farming could be widespread using vast solar-powered brewing facilities to produce edible nutritious protein.

# Conclusion

Since discovering the world of gut microbes I am constantly reminding myself to view medicine differently to the way I was taught. We now know that gut microbes are inseparable to our immune system and linked to virtually every modern disease we face, as well as fighting ageing. We also know that our microbes' health determines how we respond to medications such as painkillers, antidepressants or anti-cancer drugs. Having a healthy, diverse gut microbiome with more good bugs than bad bugs is essential for optimal health.

Yet the medical establishment has been slow to grasp the initiative. There is still little funding for nutrition studies or exploring gut-friendly diets and it has been left to a handful of academics and companies to keep pushing this agenda. Medical students today still learn little about the gut microbiome. Only a few countries, including Kenya, South Africa, Australia, India, Sri Lanka, Oman, Qatar and Bulgaria, have so far included recommendations on gut health and ferments in their national dietary guidelines, and it is about time others caught up.

## Fermentation and health

In my other books and public talks, I've identified five simple messages for gut health worth remembering, of which eating ferments was one.

1. Eat a diverse range of plants – ideally thirty different plants a week.
2. Eat the rainbow – plenty of colours to include high polyphenol foods.

3. Eat regular fermented foods – ideally three portions a day.
4. Try to practise time-restricted eating (with a 10–12 hour eating window) and avoid snacking, especially at night.
5. Reduce your intake of high-risk processed foods.

All of these will improve your gut health and the more you can do the better – ideally all of them. I'm often asked how they rank against each other. The answer is it depends on the individual and the starting point. In general, I would put fermented foods second on the list after improving plant diversity, as you need to feed your microbes properly as a baseline. Eating more fermented foods also means that you are likely to replace ultra-processed foods, which are harmful for your gut microbes and gut lining and make you overeat. Cutting out all UPFs is really hard, as around 60 per cent of our food in the UK and US comes in this form. But they are not all bad, and even reducing this to the average levels of healthier countries (10–20 per cent of calories) will give benefits. At ZOE we have created a novel food processing risk score, which breaks down the single UPF category into four risk groups. This is now available on an app in the US and we are rolling it out to other countries. Everyone should avoid the worst high-risk category. If you can get regular ferments into your daily diet, you will improve both the number and function of your beneficial bugs and reduce the numbers of bad bugs. As we have seen, this will have good effects in the short and long term. So whether you or members of your family suffer from allergies, IBS, autoimmune disease, depression or anxiety, or are taking antibiotics or have a higher risk of cancer, you should be trying to add more fermented foods to your diet, regardless of whether you also require medication.

While most countries are relaxed about fermented foods across the lifespan, a few raise concerns about young infants eating fermented foods when the gut microbiome and immune system are less stable. It takes at least three years for the gut microbiome to stabilise, although in Korea infants are given kimchi as soon as they wean, and many other countries encourage early eating of yogurts, so they are unlikely to cause harm in healthy infants. At the other end of life, it could be even more important to take regular ferments to slow down ageing and dementia, which are often related to inflammation.

Although they are safe, this does not mean they are magic bullets for everyone. Trials of people regularly eating fermented foods, as well as the results of the world's largest ZOE Ferment study of 10,000 people, still show that not all people respond well. These differences are probably related to our unique set of gut microbes. For those 6,493 people in the ZOE study that did significantly increase their ferments, most people benefited and only a few didn't, and we couldn't see any clear differences in responders between people of different ages, or body weights or by gender. The one exception was that the greater the weight of the volunteers, the more benefit they saw from a reduction in hunger.

As we study bigger numbers and build massive databases, providing personalised advice is likely to solve this problem of a few people having negative responses to ferments, as is changing to a different set of fermented foods or probiotics. But the overriding message here is that whether you are male or female, old or young, overweight or not, you are highly likely to benefit from eating more ferments.

## Adding ferments to your daily life

My journey with fermented foods since 2011 has taken me a long way; from a place where I thought eating a soy milk low-fat yogurt was good for me, to where I am now, surrounded with ferments in all shapes and sizes. I have had to move books to give me shelf space for some of my long-term ferments like miso pastes. I also had to invest in a bigger fridge, although my wife complains I have filled it up and need a second one. My double fridge is full of jars of all sizes with starters, kefir grains, scobies, vinegar mothers and bottles of kombucha and tibicos, as well as milk kefir. I try to have a Kilner jar full of kimchi and another with kraut, and of course I couldn't be without cheese. On top of this I often get sent extra commercial brands to try out. My personal and my work life have combined as I try to introduce more ferments into the ZOE program and try to develop novel prebiotics and personalised probiotics. As someone with minimal kitchen skills beforehand, in no way would I have dreamt I would be advising anyone on home cooking. I have also been amazed at how easy fermenting is, as well as how easy it has been to increase consumption compared to my preconceptions. I learned there are only a few simple rules to follow, mostly around using clean equipment and adding the right amount of salt or sugar, and the rest is fun trial and error as you personalise your own inventions.

An ideal day this summer (as I'm writing this) involves me having an early morning swim, then preparing the next batch of rye sourdough bread by feeding my starter with some flour and water. I will then check on my other ferments, such as my kombucha, which today is nearly ready to be bottled for a short second fermentation with some fruit. At about eleven, after doing some more writing I will prepare my late breakfast or brunch, which is usually a mix of full-fat yogurt, homemade kefir, peaches, apricots and my seed and nut mix of over ten varieties, washed down with black coffee. For my late lunch I will have some rye sourdough, raw milk goat's cheese (from the Pyrenees) with red cabbage and apple kraut and some lemon and ginger turmeric tibicos to drink. If it's hot during the day I'll sip on a big glass of my peach kombucha with plenty of ice. For the evening meal I'll start with a tomato and spring onion salad with olive oil and my own apple cider vinegar dressing. Then I'll have a diverse vegetable curry, mixing in at the end my milk kefir and turmeric spice mix and serving with my pepper chilli and garlic relish. I'll finish with some more seasonal fruits, raw milk cheese with kimchi on the side and, of course, a glass of red wine.

Of course every day is different, and days like this one sadly don't happen regularly enough. If you travel, have a young family or have a social life, it's impossible to be in total control, which means you need to do more planning of where your ferments will come from. As an example of an all-too-common non-ideal day, I was in LA recently for work and the hotel breakfast menu was traditional American (eggs, bagels, waffles, etc.) and offered no fermented foods save for a low-fat, artificially sweetened highly processed yogurt. So I just had the black coffee, found a store nearby that sold kefir and kombucha and took a swig, then put them in my minibar fridge. The next day I ordered the berries and black coffee and added my own milk kefir and used a bag of the prebiotic ZOE Daily30+ that I now pack with me. Sometimes when travelling shorter distances I take a tablespoon of my own kefir kept in a small make-up jar or ziplock plastic and add it to some locally bought milk when I arrive. Travelling with your own kimchi in a Tupperware container may be a healthy option, but after a few accidents I found it too risky for my laundry and friendships. Much more acceptable is to bring a few bottles of kombucha, kefir or other ferments as great practical presents for friends. This works well to both help yourself and help spread the word.

While I try to eat three different ferments per day, I know that luckily no harm is done if we don't feed our microbes for a day or two; they have evolved to be very robust. It is important to not get over anxious or obsessional about it and I realise this is a new concept for many people. As every year passes fermentation gains in popularity, and so will its availability in shops, restaurants and in people's homes, as we become increasingly familiar with the 'crazy' idea of feeding your gut microbes.

Even if you don't have the time, kitchen facilities or patience to ferment at this moment, you can buy most ferments you need if you shop around, either locally or online. I hope this book has helped you know what to look for. Generally avoid products with too much sugar or artificial sweeteners or other flavourings that are used as a lazy way to disguise the taste. You should also avoid low-fat products that are a clear sign that the product has been tampered with and replaced with cheaper chemicals and additives. While some products from trusted companies with added fruits may possibly be ok, others are not and it is always better to add your own fruits, but this will depend on your circumstances, and whether you are buying something on the go.

### *Don't be too fussy*

Depending on where you live or shop, your options may be limited. Remember, whenever you buy a fermented food to eat or drink it is likely to be a positive choice and is usually replacing something else less healthy. A supermarket blueberry flavoured yogurt is still healthier than supermarket bread and jam or most breakfast cereals; a sugary kefir is better than any children's yogurt or fromage frais; eating a vinegar-killed sauerkraut is still generally a much healthier option compared to eating chips, a protein bar or iceberg lettuce. When buying cheeses, don't get obsessed with raw milk cheeses to the extent that you avoid some very good pasteurised cheeses that are both tasty and healthy, and while you should avoid ultra-processed cheese slices that have shelf lives running to years, most cream cheese spreads still have live microbes, and will be a healthier option than chocolate milk or spreads. If you can't find the expensive apple cider vinegar, don't despair, buy a cheaper version or a wine vinegar instead, so you can still have your acidic salad before the main meal. Most microbes, dead or alive, are good microbes.

### *Try new ferments*

The first time I tried kimchi or Swedish fermented milks, I thought they were horrible, yet now I love them. Much like my first beer as a teenager. We have to realise that not only do our tastes change over time, but we can easily train our palates if we want to. As children we all had to educate our taste buds as we were weaned off sweet breast milk, and parents often have to keep reintroducing a new food to an infant until it is accepted. Korean parents have treasured videos of their baby's first kimchi – some with smiles, some with grimaces – but they all eventually love it. Many of the fermented foods have strong smells and flavours and often acidity. If you have spent the first forty years or more of your life never encountering these strange tastes and smells it is not likely you will immediately fall in love with them, like you would a doughnut. The solution is patience. Try only half a teaspoon first and try to analyse the smells and sour flavours. Dilution is initially helpful. For kombucha you can dilute it with water or if you find it very sour, add some lemonade until you find you can handle it. For kefir, add it to plain or flavoured yogurt, for kimchi, add it to cream cheese, gradually increasing the amounts and proportions. This process can take months.

### *Be flexible*

When I was seven years old my father was invited to a medical conference in Tokyo and my brother and I went with him. We were the only children there and we had six very attentive Japanese ladies in kimonos looking after these cute western boys full-time. They filled our days with origami and chopstick lessons – and Japanese food. I still remember the sushi and octopus (with suckers) covered with miso to this day. (I can also still count to one hundred in Japanese, which apart from being a good party trick, hasn't been very useful yet.) They were so delighted to see us eating well that, being polite children, we couldn't say we didn't like it, so we managed to chew it, while smiling nicely, before spitting it out discreetly into a nearby flower pot. This meant that it took me a long while to start liking Japanese food again. But now I love it. Most westerners find fermented natto one of the toughest foods to enjoy, perhaps because of its sliminess and appearance, but I now seek it out for breakfast. Asians often find smelly soft cheeses disgusting for the same reasons. It is just a question of getting used to them. I was filming recently with the TV presenter Davina McCall who I have got to know well; she said she couldn't drink kombucha and disliked sour ferments. I gave her a taste of an elderflower and ginger version from Momo and she loved it. I have now got her and her family converted to kefir and even kimchi. So often it is about staying flexible, not saying *I don't like fermented food* but instead saying *until now I haven't liked any I have tasted*. As the range of traditional and novel fermented foods continues to expand, we need to stay flexible and always be prepared to try new ones.

Happy fermenting!

### *Tim's ten top fermentation tips*

1. Eat fermented foods daily, ideally three servings a day for optimal health benefits.
2. Fermented foods contain up to fifty species of microbe, so try and eat a wide variety.
3. Fermented foods reduce inflammation, help boost immune cells, improve mental health and metabolism and can help fight cancer and the effects of ageing.
4. Try to taste different types of fermented foods to see which ones you like initially and which ones still need to grow on you.
5. Fermenting at home is cheap and much easier than you might think if you follow just a few simple principles.
6. Fermenting is safe if you follow basic hygiene and ensure sufficient salt and sugar to maintain acidity, and control the amount of exposure to air.
7. Fermenting is a great way to reduce food waste and preserve food for longer.
8. Cooked or pasteurised fermented foods like sourdough bread, traditional beers or long-life kombuchas could have some residual health benefits due to the dead microbes or their chemical products (postbiotics).
9. Precision fermentation, if it receives support, could mean that most milk, cheese, fish and meat protein can be made by microbes without killing billions of animals, and help save the planet.
10. Precision fermentation and probiotic and postbiotic mixes will expand to produce totally novel food products as well as vital novel medicines and anti-cancer drugs.

PART FIVE

# Fermenting at Home

## Getting started

Think of home fermentation as like setting off on a nice walk along the beach: rushing is futile. Since I took my first steps many years ago, there has been plenty of trial and error, and many surprises, but I have always been motivated by the fascinating alchemy that transforms mundane ingredients into culinary wonders. The basic elements of fermentation are surprisingly simple. You don't need to set up a science laboratory or buy special equipment. With a few simple ingredients, and some practical tips, anyone can try fermenting at home.

### Ingredients

When it comes to ingredients, you don't need an exhaustive grocery list, and sometimes as few as two ingredients will be enough to create a ferment. Two components are essential: firstly, a fermenting 'base' and secondly, a sprinkle of salt or kefir grains (more on these later). A base could be vegetables containing natural starter microbes for wild fermentation – where you rely on microbes already naturally present on the produce – to make a sauerkraut, or fresh milk as a base from which to grow your starter of kefir or yogurt.

For vegetable or fruit ferments, select fresh, firm plants free of obvious cuts, bruises and blemishes, which could mean some 'bad' or unhelpful microbes have got inside. Try to use quality, non-iodized mineral salt. For kefir, source fresh (ideally organic) milk and active kefir grains where possible (i.e. still alive). Think of it as creating an enticing playground for the tiny microbes, ensuring they have the right nourishment and environment to work their magic and avoid giving the microbes inferior produce to work with that is already past its best.

## Salt

Salt, in combination with sugar and acidity, is the harmonious triad underpinning getting the environment perfect for fermentation. It is important to add the correct quantity of salt: too much and it becomes inedible, too little and other bugs can infiltrate and ruin the party. Salt also allows different complex flavours to develop. By rubbing salt onto vegetables, osmosis occurs and water is drawn out as the salt is taken in, until the amount of salt on the outside and inside is balanced. What this means is that salt not only protects the food, it adds flavour and alters its texture, keeping it crisp if used wisely.

For self-brining vegetable ferments (such as sauerkraut), where you only add salt not water, the standard amount I use is 2 per cent of the weight of the vegetable, though other recipes might suggest a bit more or less. For brine fermenting (where a brine made of water and salt is added to whole or large vegetables without massaging them, you should use a bit more salt – say 3–4 per cent. So for 400g (prepared weight) of vegetables, 2 per cent salt for self-brining would be 8g (2 teaspoons). Note that if you need to top up the jar with 200ml water, you will need to add another 8g, giving a total of 16g (4 per cent) salt in this example. If the vegetables just need a small top-up, it's fine to weigh the total contents on digital scales and add 2 per cent salt. If you are in a hurry, a tablespoon of salt is fine for a kilogram of vegetables. Make sure the salt is evenly spread over the plant, either by rubbing or mixing the briny water around. Sometimes you can add extra salt at the top of the jar or under the lid where it is most vulnerable to microbial competition. Some recipes suggest adding slightly more salt in summer to slow the process down a bit. The longer the ferment, the more salt you need to protect it.

I prefer to always use sea salt rather than table salt as it has more minerals, including calcium, magnesium and potassium, which provide extra nutrients for the fermenting microbes and give the foods extra depth of flavour. Another reason is because of anti-caking agents found in most table salts, most of which are 99 per cent pure refined sodium chloride and lacking any interesting tastes. I have found that ordinary table salt still works in an emergency, although the taste is harsher and salt with iodine will inhibit many bacteria. Start with what you have, but experiment with higher quality mineral salts as you gain confidence. It's important to note that if you are using sea salt, coarse (kosher) salt or table salt, their granules are very different sizes, so table salt will always be saltier if you go by teaspoons not grams. If in doubt, do weigh the salt – digital scales can be helpful here.

### *Acidity*

Acidity is brought about by the industrious microbes that flourish in the salty environment and is the counterpoint that provides balance and depth to the flavours. Just as an orchestra is incomplete without the soul-stirring notes of the violin, a ferment without acidity lacks character and vibrance. In food, just as in our guts, tiny changes in acidity mean big changes in the microbes and how they proliferate. Acidity – and the opposite alkalinity – is measured on a pH scale from 0–14 where a pH of 7 is neutral, 1 is super acidic and 14 is super alkaline. Most harmful microbes do not like living in acidic conditions. However, lactic acid bacteria thrive there and, as they produce lactic acid, they further reduce the pH to below the magic 4.5 within a few days. This stops passing pathogenic bugs like *Listeria* from being able to survive in the ferment, keeping it safe. Most of my ferments drop below a pH of 4 rapidly within three days. I bought a pH meter on the internet pretty cheaply and found it both fascinating and comforting to check the acidity and therefore safety of the new concoctions I was making. You can also buy pH strips of litmus paper, which are even cheaper and easier to use, though can be tricky if your ferment is brightly coloured like kimchi.

### *Sugar*

This is commonly added to fermented drinks as a source of energy for the yeast or other microbes. As with salt, basic refined table sugar will work but is not ideal as it may contain anti-caking agents, so do check the labels. Raw organic cane sugar is probably the most versatile sugar for fermenting, especially for water kefir and kombucha, which have cultures that can be sensitive to excess minerals. Other sugars richer in molasses and minerals (brown, muscovado, rapadura, etc.) can be used and add both colour and extra flavour but can sometimes interfere with kefir grains. Adding a small amount of molasses or maple syrup to other sugars can add extra flavour but are tricky to use on their own. Avoid coconut palm sugars, artificial sweeteners, stevia or agave. Honey, ideally raw and unpasteurised, can be used as fuel for fermenting microbes, and is activated if diluted by just 20 per cent. This is the basis of the simplest fermenting recipe in this book – honey-fermented garlic (see page 240).

### *Water*

Water varies widely depending on where you live, with hard water (containing calcium and phosphate) generally being better for fermenting (although worse for your hair). If you have soft water some recipes suggest adding fruits such as dates to kick-start a ferment. Ideally you should use filtered water, left to stand for a few minutes to allow the chlorine to dissipate, although I sometimes forget and it still works. Microbes are remarkably robust and chlorine dissipates rapidly into the air. If you are unsure and don't have a water filter, boiling and cooling tap water is a failsafe trick, unless you want to buy bottled water with its large carbon footprint.

*

Home fermentation can also require a few other more specialised ingredients. Fermentation 'starters', such as mother cultures or grains, can be sourced online, from health food stores, or even borrowed from fellow fermenters.

## Safety and hygiene

The symbiotic dance of microbes during fermentation also involves moulds and yeasts, but it's important to distinguish between beneficial and harmful ones. Just like in most neighbourhoods, undesirable elements occasionally emerge, but their presence is usually noticeable. Bright unexpected colours such as pink or purple, or black or dark brown as well as foul odours are red flags, indicating that something has gone wrong. Luckily, most moulds are opportunistic surface dwellers; the commonest is a harmless kahm yeast, which is white and produces a thin layer that can easily be scraped off. I've included other tips on spotting danger in the troubleshooting section (see page 267).

Wash your hands, equipment and work surfaces before you start and use clean spoons each time you check on your ferment, but other than that there's no need to sterilise or boil everything. Most microbes such as *Listeria* can't live above 65°C, the average temperature of hot water from a tap; remember, we're creating a live theatre, not conducting open-heart surgery.

## Environment

Air, an indispensable element of life, is key to the success or failure of your fermentation drama. For sauerkraut and other lacto-fermented vegetables, it's crucial to limit exposure to air as we're favouring anaerobic bacteria (bacteria that flourish in a lack of oxygen). This can be achieved using plastic lids, ziplock bags filled with salty water, dry rice or chickpeas or weights (commercial glass weights or boiled cleaned round stones) to keep the produce submerged in the brine, away from the air. Or I simply often use the tough outer leaves of a cabbage as the surface protectors. However, for ferments relying on yeast, such as sourdough and kombucha, as well as vinegar and some kefirs, access to air is crucial. Generally, try to avoid wide open lids as they attract fruit flies and other insects, so cover them with a muslin cloth; I've found that fine kitchen paper towels with an elastic band work fine too.

## Equipment

You don't need a science lab to ferment at home, and most people will already have containers they can use in their kitchens. Non-reactive materials such as glass, ceramic or food-grade plastic are ideal. Metal containers, except for high-grade stainless steel, are best avoided as they may react with the acids produced during fermentation, leading to off-flavours or potential toxicity. The microbes on kefir grains are especially allergic to metal instruments, so use plastic or wood for spoons or sieves if you can, or at least limit their exposure, by covering the metal with cloth or paper. I tend to use old glass jars of different sizes or Kilner type jars with a clip lid and a rubber washer that allows some gas to escape if it builds up. Having a few sizes of funnel is useful for decanting into bottles, as is a coffee filter or kitchen paper, which I use to drain out the water if concentrating a sauce or yogurt or making a cream cheese. If making kombucha or fruit kefirs with a second fermentation, I use a plastic bottle for some of the batch, so I can tell how much gas is being produced. An over-frothing or exploding bottle after a second fermentation is one of the major risks of fermenting – you have been warned!

A temperature gauge is crucial for yogurt, and pH meters are a handy accessory when you are starting out, as are digital scales you can set to zero to weigh your vegetables and salt. When fermenting in a glass jar or similar vessel, a general guideline is to leave at least one inch of headspace to allow for the expansion that occurs during fermentation. This also helps prevent brine overflow, ensuring a cleaner and more controlled process. In any case, you should always place something underneath your jars as there tends to be some bubbling and leakage, and as I have discovered red cabbage sauerkraut leaves quite a purple mess. Vacuum-packing machines are now cheap and accessible and putting your ferments in them is a safe, controlled way that will become more popular and worth trying.

## Time

Time and patience are crucial. The length of fermentation is a key determinant in the sourness of the final product. Longer fermentation periods allow the *Lactobacillus* to produce more lactic acid, resulting in a more pronounced sourness, or a yeast to use more of the sugar.

The duration of fermentation can vary a lot depending on ambient temperature and personal taste preference. Most ferments and palates prefer slower, longer ferments at cooler temperatures. This is a game of patience and observation, as the balance can shift from bland to tangy to overly sour. Sample your ferment periodically to find your 'sweet spot' for sourness and texture, or by simply looking closely to see how actively it is bubbling. Once your ferment has reached the desired flavour and acidity, it can be moved to the fridge to slow down the fermentation process to glacial speed, effectively hitting the 'pause' button.

## Other practical tips

When fermenting something new I start with small batches so I can fine-tune the process and get to understand the rhythm of fermentation. No one likes to throw litres of milk or four cabbages away in a failed first-time ferment.

Make sure that you are going to be able to check on your new fermenting pets and will be around to burp them or stir them for the first few days. Kefir and sauerkraut are excellent quick beginner's projects – they are forgiving and yield flavourful results in a few days.

And most importantly, embrace the unexpected! Fermentation is a naturally varying process – each batch tells its own tale and unfolds its unique melody. Don't forget to take notes and label your jars with dates and ingredients, so you can learn from mistakes and successes. I now keep a fermenting notebook (and I wish I had done that from the beginning).

Starting your first batch of 'cold cooking' might seem daunting, but the magic of fermentation is in its natural variability. It's as much an art as an exact science, so we can be liberated from the strict confines of recipes. Trust your senses and intuition and remember that the risk of food poisoning is tiny if you follow good hygiene practices and keep an eye out for signs of spoilage. The fermenting journey is full of surprises and learnings, leading you to a destination of profound flavours and enhanced health. Go for it. And even if you're not quite ready to try fermenting yourself now, it's helpful to understand what goes into some of the most common and basic ferments, and how you can incorporate them into your diet.

### *Getting started in five*

1. Basic hygiene with soap and water is important, but there's no need for obsessional sterilisation.
2. You can start with any old glass jars, plastic tubs or bottles with lids, and add more containers and equipment as you need them.
3. In ferments that require water, ideally use filtered water, or make sure you leave tap water for a while before adding it to a ferment (to allow the chlorine to dissipate).
4. Quality milk, salt and sugar will enhance your ferments, due to the extra minerals and lack of additives.
5. A set of digital scales is very useful, a thermometer and a pH meter or litmus paper are nice to have.

# Recipes

## Yogurt

The basic formula is much the same, whether you are making dairy or non-dairy yogurt. For non-dairy options soya and coconut milk work best. You will need clean jars with lids or a wide-necked thermos flask.

### Dairy yogurt

MAKES 1 LITRE

*1 litre organic whole milk*
*2–3 tbsp good-quality natural yogurt (with no additives), at room temperature*

Slowly heat the milk in a saucepan to 85–89°C (or as close as you can without boiling or burning it). Stirring often, keep the milk at this temperature for about 5–10 minutes to kill all bugs and break down the casein protein.

Remove the pan from the heat and leave the milk to cool to 45°C. Meanwhile fill the jars or thermos flask with freshly boiled water, leave for 2–3 minutes to warm up and then drain thoroughly.

Add the yogurt to the warm milk and mix to combine, then pour the milk mixture into the warmed jars, cover with lids and keep somewhere warm – if possible above 38°C – and do not touch for 8–12 hours. The jars can be wrapped in towels and kept in an airing cupboard or above a warm radiator. Alternatively pour the mixture into a wide-necked thermos flask, seal the lid tightly and leave at room temperature for 8–12 hours.

Check the yogurt has set and thickened and store in the fridge. Keep some of this yogurt to make the next batch.

It's that easy.

---

## Soya yogurt

MAKES 1 LITRE

*1 litre good-quality soya milk*
*2 tbsp shop-bought natural soya yogurt, at room temperature*

In a pan heat the soya milk to 40–42°C.

Remove from the heat, add the soya yogurt and stir to combine. Pour into warmed jars, cover and keep somewhere warm at around 38°C for 8–12 hours until set. Refrigerate until ready to use.

---

## Coconut yogurt

MAKES 400ML

*400ml tin organic coconut milk (without additives, thickeners or guar gum)*
*2 tbsp good-quality natural organic coconut yogurt, at room temperature (or 2 probiotic capsules)*

Tip the coconut milk into a jug and whisk until smooth. Add either the yogurt or the contents of the probiotic capsules and stir to thoroughly combine.

Pour into a jar, cover with a cloth secured with a rubber band and leave at warm room temperature for 12–24 hours until thickened and tangy.

Chill the yogurt to thicken further before serving.

## Cultured Butter

This rich and luxurious version, made from fermented cream rather than plain cream as for normal butter, is tangy and complex in flavour.

MAKES ABOUT 325G

*600ml double cream (with no added thickeners or emulsifiers)*
*60–90ml (4–6 tbsp) natural yogurt or buttermilk*
*Sea salt flakes*

Combine the double cream and yogurt or buttermilk in a clean glass jar or ceramic or glass bowl. Cover and leave at room temperature (but out of direct sunlight) for 24–36 hours, by which time the cream will have thickened.

Chill the mixture for 2 hours.

Tip the cream mixture into the bowl of a stand mixer fitted with a paddle attachment and mix on lowish speed for 2–3 minutes until the cream becomes very thick and then separates into thick yellow butter, which will collect around the paddle with the liquid buttermilk sitting in the bowl below.

Tip the solid butter and buttermilk into a sieve set over a bowl. Bottle and chill the strained buttermilk for another use and then use your hands to massage and squeeze the solid butter under very cold running water to rinse off any excess buttermilk. Return to the clean mixer bowl, add a good pinch of sea salt flakes and mix again to combine.

Shape the butter into pats or logs, wrap in greaseproof or waxed paper and chill until ready to use.

## Cream Cheese/Labneh

This is probably the easiest cheese that you can make yourself and is great if you have an excess of kefir or yogurt.

MAKES AS MUCH AS YOU LIKE
*Large pot of homemade or shop-bought kefir or yogurt*

OPTIONAL EXTRAS
*Salt and freshly ground black pepper*
*Fresh herbs*
*Lemon zest*
*Spices*
*Olive oil*
*Garlic*
*Chillies*

Take your kefir or yogurt and tip it into a sieve lined with a coffee filter, cheesecloth or new clean J cloth. Sit the sieve over a bowl and leave the yogurt or kefir to drip overnight in the fridge. The probiotic whey that drips out of the yogurt or kefir and collects in the bowl can be used as a starter in another recipe or to add extra protein to other dishes. If the strained cheese is very wet still it can be strained a second time.

Either use the cream cheese as it is or season with salt and freshly ground black pepper and add fresh herbs, lemon zest and/or spices. I love to add chopped chives and crushed garlic or a ras el hanout spice mix and then serve with sourdough bread.

If your cream cheese is nice and firm you can also roll it into neat balls and marinade in olive oil with garlic, chillies and fresh herbs such as thyme, rosemary, parsley, basil or oregano. This is a great addition to salads.

## Simple Sauerkraut

Traditionally sauerkraut is fermented slowly at lowish temperatures – 2–4 weeks at around 15–20°C – to give enough time for the sourness to properly develop. With a pH of 3.5 sauerkraut is around seven times more sour than kimchi, its spicier cousin. Red cabbage, takes longer to ferment than white cabbage, and both work faster if you add another vegetable (like a small carrot or two) that contains a range of accessible sugars for the microbes.

MAKES 1 LARGE JAR
*1 white or red cabbage*
*1–2 carrots (optional)*
*1 tbsp caraway seeds (or juniper berries)*
*Sea salt*

Trim the base of the cabbage and peel away the outer leaves; set these aside for use later. Cut the cabbage into quarters and use a mandoline, food processor, coarse grater or sharp knife to finely shred the cabbage and carrot (if using).

Tip the cabbage into a large bowl, add the caraway seeds and make a note of the total weight. Add 2 per cent salt of this total weight, so if the total weight is 600g, you will need 12g salt.

Using your hands, massage the salt really well into the shredded cabbage for a couple of minutes until it starts to soften. Cover the bowl with a clean cloth or plate and set aside for at least 30 minutes and up to 3 hours until the cabbage is very soft and has released water.

Scoop the cabbage and any resulting liquid into a clean 1–2 litre jar and really pack it down hard so that the shredded cabbage is submerged under the briny liquid and you have a clear 5cm gap between the cabbage and the top of the jar. Cover the top of the cabbage with the reserved outer leaves and place a weight (or the cabbage core) on top. Close the lid but leave it slightly loose – if the jar is sealed tight it runs the risk of exploding!

Place the jar on a plate in a dark cool cupboard out of direct sunlight and leave for 2–3 days for fermentation to start. Burp the jar daily to release any collected gas. After 7 days the sauerkraut should be fermenting nicely so it can now be stored in the fridge. Use within 1–2 months, depending on how soft you like it.

## Traditional Kimchi

In this traditional kimchi recipe (one of many variants) the cabbage is salted first; after a couple of hours it is rinsed off and then the cabbage is rubbed with the spicy kimchi paste of chilli flakes, glutinous rice powder, garlic and ginger.

MAKES 1 LARGE JAR

*2 Chinese/napa cabbages*
*2–3 tbsp table salt*
*20g glutinous rice flour*
*250ml water*
*¼ daikon, peeled and cut into matchsticks*
*1 carrot, peeled and cut into matchsticks*
*5–6 spring onions, thinly sliced*
*1 onion, roughly chopped*
*8 fat garlic cloves, peeled and chopped*
*Thumb-sized piece of fresh ginger, peeled and chopped*
*3 tbsp fish sauce*
*50–75g Korean chilli flakes (gochugaru)*
*1 tbsp golden caster sugar*

Start by trimming any bruised outer leaves from the Chinese cabbage and trimming the base. Cut each cabbage into quarters through the root. Sprinkle the salt liberally all over the cabbage and thoroughly in between each leaf and place in a large bowl. Set aside for 2 hours to soften, turning the cabbage once or twice during this time.

Meanwhile prepare the kimchi paste. Whisk the rice flour with the water in a small saucepan until smooth and set over a low-medium heat. Bring to the boil, reduce to a low simmer and cook for 3–4 minutes, stirring often until translucent and thickened. Remove from the heat and leave to cool.

Put the daikon, carrot and spring onions into a large bowl. Tip the onion, garlic and ginger into the bowl of a food processor with the fish sauce, chilli flakes and sugar and blend until smooth. Add this paste to the carrot and daikon along with the cooled rice flour paste and mix to combine.

Thoroughly wash the salt from the Chinese cabbage under cold running water and leave to drain well in a colander for 10 minutes or so.

Using your hands (clean rubber gloves are useful here), massage the chilli and vegetable paste into the cabbage, ensuring that each leaf is well coated. Fold the quarters in half into neat parcels and pack tightly into a plastic food container. Cover with a lid and leave at room temperature for about 48 hours until the kimchi starts to ferment. During this time you can press the cabbage down into any liquid that forms.

Once the kimchi has fermented for a few days it should be stored in the fridge where it will keep for weeks.

## Simple Kimchi

This is a short cut for those who want a faster recipe, cutting out the traditional two-stage salting process (see recipe on page 233). Try both methods and see if you can tell the difference. You can use any whole cabbage, Chinese/napa cabbage and/or daikon (Japanese radish) for this recipe.

MAKES 1 JAR
*1 cabbage or Chinese/napa cabbage*
*1 daikon*
*Sea salt*
*4 garlic cloves*
*1 bunch of spring onions, trimmed and sliced*
*1–2 tbsp Korean chilli flakes (gochugaru)*
*1 tbsp soy sauce*
*1 tbsp fish sauce (or fermented miso paste for vegans)*

Trim the cabbage and cut it into 4–5mm thick slices. Trim and peel the daikon and cut into thin half-moons.

Weigh the combined vegetables, tip into a bowl and add 2 per cent salt of the total weight. Using your hands, massage the salt into the vegetables until they are starting to soften, then cover the bowl and set aside at room temperature for about 4 hours.

Combine the garlic, spring onions, chilli flakes, soy sauce and fish sauce in a food processor and pulse to combine. Add to the vegetables and mix well to thoroughly coat them. Pack tightly into a clean jar ensuring that the vegetables are submerged beneath the brine. If necessary, add a little 2 per cent brine (i.e. 2g salt for every 100ml water) to cover. Press weights on top and loosely cover with a lid, then leave in a cool, dark place for 5–10 days until fizzy. Transfer to the fridge and eat within 1–2 months.

#### VARIATION: LEFTOVER KIMCHI

Chop any spare vegetables including peppers, brassicas, carrots and onions (but avoid anything too leafy and too soft) into 4–5mm slices. Finely chop a few garlic cloves, a thumb of ginger and 2–3 chillies and add to the mix.

Weigh the prepared vegetables and add 2 per cent sea salt and a good amount of chilli flakes (Korean or ordinary) and mix to combine. Using your hands, massage the vegetables for a couple of minutes to soften and then leave in a covered bowl for an hour. Then tightly pack/squash forcibly into clean jars, making sure that the vegetables are submerged below the resulting brine. If necessary, you can top up the liquid with 2 per cent brine.

Cover loosely with a lid, sit the jar on a plate and store somewhere dark and cool for a week, burping the resulting gas every day or so. Store in the fridge and eat within 2–4 weeks.

#### WAYS TO ENJOY KRAUTS AND KIMCHI

- Add to salads or stir into rice or grain dishes – kimchi is particularly delicious in egg fried rice or served with noodles.
- Add to cream cheese as a spread – a great idea for kids.
- Add sauerkraut to a Reuben sandwich with pastrami, salt beef or mushroom pate on rye bread.
- With cheese and bread or in toasted cheese sandwiches.
- Add to dumplings or macaroni cheese or serve alongside scrambled or fried eggs.

## Kimchi Bloody Mary

The complexity in flavours in this iconic cocktail-with-a-twist comes from the ferment. By first blitzing the ingredients (including the whole kimchi) and then straining the liquid, you get an intense flavour, plus a dose of live microbes.

SERVES 4–6

*250g kimchi, including the brine*
*250ml vodka*
*1 litre tomato juice*
*A squeeze of lemon juice*
*1 tbsp horseradish*
*Sugar to taste*
*Celery salt to taste (optional)*
*Worcestershire sauce to taste*

TO GARNISH

*Ice cubes*
*Pieces of kimchi*
*Green olives*
*Lemon slices*
*Gochugaru*
*Celery sticks*

Put the kimchi, vodka, tomato juice, lemon juice and horseradish into a blender and blitz until nearly smooth. Pass through a sieve into a clean jug, then taste and adjust the seasonings, adding a little sugar, celery salt and Worcestershire sauce to taste.

Pour into ice-filled highball glasses and add your garnishes. Skewer a piece of kimchi and an olive onto a cocktail stick, add a slice of lemon and a sprinkle of gochugaru and serve with a celery stick.

## Tepache

This is a refreshing drink from Mexico where piloncillo sugar is used, but any organic unrefined sugar will work. It's an easy recipe that requires no specialist equipment other than a 3-litre Kilner jar or similar.

Once ready to drink serve tepache in tall glasses over ice with a slice of lime.

MAKES 1.5 LITRES
*1 large ripe (ideally organic) pineapple*
*Thumb-sized piece of fresh ginger, peeled and thinly sliced (optional)*
*2 cinnamon sticks (optional)*
*150g (5 per cent) unrefined cane or brown sugar*
*1.5 litres water*

Trim the leafy top and bottom stalk end from the pineapple and discard. Wash the pineapple and stand it upright on a chopping board. Cut away the skin in long downward strips and roughly chop, then add to a clean 3-litre jar.

Cut the pineapple into quarters and slice away the core. Roughly chop the core and add to the jar, along with the ginger and the cinnamon sticks (if using). The pineapple flesh can now be eaten or used in another recipe.

Stir the sugar into the measured water until dissolved, then pour into the jar to cover the pineapple skins. Cover the jar with a clean cloth secured with string or an elastic band. Leave at room temperature (but out of direct sunlight) for 2–3 days until you can see bubbles appearing.

Strain the liquid, decant into bottles and chill before serving. This will keep in the fridge for 3–4 days but if left too long the tepache will become alcoholic and then turn into vinegar.

VARIATION: instead of cinnamon sticks you can add other whole spices such as cloves, star anise, cardamom and allspice and even a little fresh red chilli for a kick. Try adding the (washed) skins of 2 mangoes to add a little more flavour.

## Honey-fermented Garlic

This is probably the easiest vegetable ferment you can make – garlic comes complete with its own microbiome starter and chemicals such as allicin, which deters other microbes. You will need raw honey for this recipe – honey that has been extracted without the use of heat and is unpasteurised, untreated and unprocessed and retains all the nutrients

MAKES 1 JAR
*2–3 heads of garlic*
*About 300g raw honey*

Peel enough garlic cloves to fill a clean jam jar or Kilner jar two thirds full. Spoon over the raw honey to cover. Close the lid and give the jar a good shake to combine the honey and garlic. Sit the jar on a plate and store in a larder or cupboard at room temperature and out of direct light. Turn the jar and 'burp' the lid every few days for 7–14 days and watch the garlic ferment and transform the thick honey into a clearer, runnier liquid. The garlic may discolour, but don't worry – this is normal.

Once the garlic has fermented it will keep for months in a cool place or the fridge. You can then use either the garlic or honey separately as stand-alone ingredients or try blending them together. Add some good olive oil and apple cider vinegar to make an amazing salad dressing. The honey can be used as a marinade, a glaze for meat, fish or vegetables or do as I often do and nibble on a whole honeyed garlic clove, which in some cultures is believed to have health benefits such as warding off colds.

VARIATION: try adding other herbs too such as sliced fresh turmeric root, ginger or a sprig of lavender to the infusion.

## Tomato Ketchup

A fermented version that has none of the additives and high sugar content of shop-bought varieties.

MAKES: 1 BOTTLE

*200g tomato purée*
*1 tbsp Worcestershire sauce or soy sauce*
*Splash of fish sauce*
*1 tbsp runny honey (you could use fermented garlic honey if you have it)*
*1 tbsp apple cider vinegar*
*2–3 tsp brine from sauerkraut (optional)*
*2 tsp coriander seeds*
*½ tsp black peppercorns*
*1 tsp mustard seeds*
*⅛–¼ tsp ground cinnamon*
*Salt to taste*

In a bowl combine the tomato purée, Worcestershire or soy sauce and a splash of fish sauce. Add the honey, vinegar and brine (if using).

In a dry frying pan over a low-medium heat, gently toast the coriander seeds, black peppercorn and mustard seeds for a minute until aromatic. Finely grind the spices with the cinnamon and a pinch of salt. Add to the tomato mixture and mix to combine.

Spoon into a clean jam jar, add the lid and keep at room temperature but out of direct sunlight. Stir every day for 3–5 days until starting to ferment.

Store in the fridge and enjoy as you would regular ketchup. Eat within 4 weeks.

## Kefir

Kefir can be made with aerobic (open) or anaerobic (closed) fermentation. Aerobic fermentation is slightly easier for beginners and creates a fizzier, more yeasty flavour because more carbon dioxide gas is produced from the yeast and other oxygen-loving microbes. Anaerobic fermentation, on the other hand, results in a sour, more constant flavour with less fizziness, which I prefer.

For anaerobic fermentation you will need a clean 500ml Kilner jar (avoid metal or plastic) and for aerobic a large glass bottle works.

MAKES 500ML
*500ml organic whole milk*
*1–2 tsp kefir grains*

Pour the milk into the jar/bottle and add the kefir grains. Leave a 2cm gap between the top of the milk and the top of the bottle.

Close the lid or cover with a cloth secured with an elastic band for anaerobic fermentation. Or for aerobic fermentation (for a frothier, fizzier kefir) simply pour into a clean glass bottle, leaving a 2cm gap at the top and leave open or the lid loose. Leave at room temperature but out of direct sunlight for 18–36 hours to ferment, gently shaking the bottle or jar from time to time.

Taste and smell the mixture regularly to check that it is fermenting and producing acid. When you are happy with the flavour and fizziness, strain the mixture through a plastic sieve into a jug, then pour into a clean container, cover and chill until ready to use. Kefir will keep for up to one week in the fridge before it starts to have a cheesy flavour.

NOTES

- The grains can be used again – do not wash them but simply place in a clean jar and start again with more milk.

- To change the flavour and fizziness in your kefir try adding slightly more grains, changing the milk or leaving it to ferment for slightly longer.
- You can use other milks such as sheep, goat and some non-dairy milks – each variety will create different flavours and fizziness profiles.

WAYS TO ENJOY KEFIR

Purists may enjoy kefir as a daily shot in a small glass but don't stop there. There are many simple uses for it which are especially useful when you have made a big batch.

- Serve kefir with a squeeze of lime or lemon as a side with curries or chillies.
- Add to smoothies in place of or in addition to milk or yogurt.
- Make into ice cream by blitzing in a blender with frozen mango or bananas with a dollop of nut butter.
- Try mango lassi – a sweet Indian favourite. Mix mango pulp with kefir to a ratio of 1 part mango to 5 parts kefir and serve chilled, perhaps with a squeeze of fresh lime.
- A favourite at my table is kefir salad dressing. Mix kefir with extra virgin olive oil, a squeeze of lemon or lime juice, crushed garlic and chopped soft herbs such as parsley, dill, chives, mint or coriander. Season well with salt and freshly ground black pepper. Try blending with a ripe avocado for a richer Green Goddess style-dressing and add to a chickpea salad for an instant flavour and health boost.
- Use kefir to make labneh in place of yogurt.

## Tibicos or water kefir

Water kefir or tibicos (also known as Timicos in my house) is one of the easiest and quickest ferments and is very hard to get wrong, making it perfect for entry level fermenting. I've yet to find anyone who doesn't enjoy this refreshing, zingy concoction and, unlike kombucha, it's caffeine-free. After the initial fermentation and bottling you can add flavours: whatever fruits or herbs are in season or your preference.

Water grains are easily available online or get some from a friend who is already brewing – the grains quickly multiply into more than you will need. It is important to use thoroughly clean but not sterilised bottles and jars and always strain through a plastic sieve and stir with either a plastic or wooden spoon – never metal. This recipe makes 500ml but can easily be scaled up after your first few batches, depending on how quickly you are drinking it.

MAKES 500ML

*500ml filtered tap or spring water*
*35–40g golden caster or granulated sugar*
*25g live water kefir grains*
*¼–½ unwaxed lemon*
*1 unsulphured dried fig or prune or 2 tsp organic raisins*
*1 slice of fresh ginger, peeled*

Pour the filtered water into a fermenting jar or Kilner jar. Add the sugar and stir with a wooden spoon to dissolve.

Add the water kefir grains, lemon, dried fruit and ginger, then cover with a cloth secured with an elastic band or fermenting lid and leave at room temperature but out of direct sunlight for 24–48 hours. How quickly the kefir ferments will be determined by the temperature of your room – the warmer the room the quicker the fermentation process.

Once the kefir is bubbly, strain through a plastic sieve and pour into clean bottles. Add any flavourings such as fresh ginger, frozen berries, elderflower, turmeric or herbs. Cover the bottles with lids and continue fermenting at room temperature for a further 24–48 hours, by which time the kefir will be nicely carbonated. Chill and drink within a couple of days.

Add more water, sugar, lemon and dried fruit to the clean fermenting jar, return the grains without rinsing and start another batch.

## Kombucha

You will need a SCOBY (symbiotic culture of bacteria and yeast) to make kombucha. These are available online, can be found at the bottom of a good bottle of shop-bought raw unflavoured kombucha or you can grow you own. Failing that, find a friend who is already brewing kombucha and have one of their babies (a SCOBY baby).

To grow your own SCOBY you will first need a small bottle (330ml or thereabouts) of raw unflavoured kombucha. Brew up 1.5 litres black tea (avoid anything aromatic such as Earl Grey) using 3–4 tea bags, add 100g sugar, stir to dissolve and leave to cool to room temperature. Add the kombucha, including any yeasty residue and strands in the bottom of the bottle, and pour into a spotlessly clean 2 litre Kilner or glass jar. Cover the jar with a double or triple thickness of clean cheesecloth secured with a rubber band or string and leave the jar out of direct sunlight and at room temperature for about 4 weeks, or until a 5mm thick SCOBY has formed. Now you can start to brew kombucha.

MAKES 1.5 LITRES
*3–4 tea bags*
*500ml freshly boiled water*
*125–150g granulated or caster sugar*
*1 litre filtered water*
SCOBY *(see above)*

Use the tea bags and boiled water to make up a brew of tea. Leave to steep for 10–15 minutes, then remove the tea bags (do not squeeze them as this will make the tea too bitter) and add the sugar and stir to dissolve. Add the filtered water.

Pour into a clean 2-litre jar and leave to cool to room temperature.

Add the SCOBY (make sure your hands are spotlessly clean when handling your live culture) and about 100ml of the original starter brew. Cover the jar with a double or triple thickness of cheesecloth and secure with a rubber band or string; this will let air in but keep insects out. Leave the jar at room temperature and out of direct sunlight and wait for 1–2 weeks as the colour lightens, and the SCOBY rises and grows to fit the container.

After 5–7 days pour a little kombucha into a glass and taste it – it should be fruity and no longer taste of tea. Continue fermenting until it has reached a level that you are happy with – anywhere from 5 days to 2 weeks is about normal.

You can now draw off the kombucha and bottle it so it is ready to drink or go for a second fermentation when you will add any flavourings and extra fizz.

Leave a cupful of kombucha in the fermenting jar and either refrigerate this until you are ready to brew again or make up more tea and immediately start again. Once you have established a brewing rhythm you can scale up the quantities you make with each brew depending on how much kombucha you drink.

To add flavours and a second fermentation pour your brewed kombucha into 500ml flip-top bottles with ¼–½ teaspoon of sugar and a choice of fruit and herbs. The bottles should be filled to within 5cm from the top to allow space for the carbon dioxide to expand and escape without liquid overflowing. Peaches, mint, lemon verbena, elderflower, pineapple, hibiscus, grenadine syrup, ginger, turmeric and berries all work well. If using only high-quality tea you can also add unbrewed tea leaves like Darjeeling to the second stage to add flavour. Seal the bottles and leave at room temperature but out of direct sunlight for 24–48 hours, keeping a close eye on the bottles to avoid too much gas build-up and exploding bottles. Chill thoroughly before serving.

### VARIATION: COFFEE KOMBUCHA

Brew a batch of kombucha and have it ready for the second fermentation. Make up a cup of fresh coffee using ground coffee – either using a cafetière, filter, cold press or however you make your coffee – and leave to cool to room temperature.

Pour about 150ml coffee into a clean flip-top bottle, add 1 teaspoon of sugar and shake to dissolve the sugar. Top up with kombucha, cover and leave at room temperature for 24 hours for second fermentation to occur. Chill before serving.

### VARIATION: KOMBUCHA CRÈME FRAÎCHE OR ICE CREAM

Add 3–4 tablespoons of kombucha to 350–500ml double cream, stir to combine and leave in a covered bowl or jar for 24–36 hours at room temperature to ferment.

Pop the cream into the fridge and chill for 4–24 hours to produce a probiotic crème fraîche.

You can now either use your crème fraîche over the next few days or pour into moulds and freeze overnight to enjoy as ice cream.

### OTHER KOMBUCHA VARIANTS

- Kombucha with fermented hops is a lovely drink, similar to beer in flavour, though hops can be hard to obtain.
- Hard kombucha is with alcohol between 0.5–8 per cent and is more commonly found in the US. Brewer's yeast is added to the kombucha at second fermentation.
- Kombucha can also be used as a starter to make cultured (fermented) butter.
- Jun is a drink similar to kombucha but is made using raw honey and green tea rather than sugar and black tea and requires a jun SCOBY. It has a milder more floral flavour, but the process is otherwise the same, and it will take less time to ferment.

## Fermented turmeric or ginger

Turmeric and ginger both have health benefits for the immune system and for reducing inflammation and pain and the effects are likely enhanced when fermented.

MAKES 1 SMALL JAR

*1 tbsp grated organic fresh turmeric or ginger (skin on)*
*1 tsp sugar*
*3 tbsp filtered water*

Combine the grated turmeric or ginger and sugar in a clean glass jar. Add the filtered water and stir until the sugar dissolves.

Cover the jar with a clean cloth or cheesecloth and secure with a rubber band. Leave at room temperature but out of direct sunlight for 24 hours.

The next day, repeat the step above adding more turmeric, sugar and water. Re-cover the jar and place back into the dark corner.

Repeat for an additional 2 days, or until the turmeric bug starts to bubble. It may take up to a week to start fermenting. Once it is bubbling it is ready to use as a starter for homemade sodas, water kefirs or as a healthy addition to food.

You can keep the bugs alive by continually feeding them in the same way as you would a sourdough starter (see page 250).

## Sourdough Starter

This is really optional for those that like to start from scratch. The process for making a starter can take anything from about 6 days to 2 weeks depending on the flour used and the ambient temperature of your kitchen. You will need a large, lidded jar and some patience.

*300g organic strong white flour*
*300g organic spelt or rye flour*
*Tepid water – tap is fine and it should be neither warm nor fully cold*

**Day 1:** Combine both flours and store in an airtight box or jar. Mix 25g of the flour mix with 25g water in a small glass or ceramic bowl and beat until smooth. Cover loosely with cling film and leave at room temperature for 24 hours. (Total weight = 50g.)

**Day 2:** Mix another 25g flour with 25g water to a smooth paste and add to the mixture from the previous day. Beat well to combine, cover loosely with cling film and set aside for 24 hours. (Total weight = 100g.)

**Day 3:** Remove half of the mixture (50g) and discard it. Add a further 25g flour and 25g water to the remaining starter and mix to combine, then scoop into a glass, ceramic or stoneware jar, cover loosely with a lid and set aside for 24 hours. (Total weight =100g.)

**Day 4:** Repeat Day 3. (Total weight = 100g.)

**Day 5:** Mix 100g flour and 100g water and add to the starter. Beat well to combine, cover loosely and leave for 24 hours. (Total weight = 300g.)

**Day 6:** Remove half of the starter (150g) and add 75g flour and 75g water. (Total weight = 300g.)

By now your starter should be bubbly and active and have a fresh yeasty smell. You can repeat this process for a further 6 days removing half of the starter and adding back the same amount of water and flour combined every day. Remember to keep the ratio of flour and water the same at every feed. If the starter is slow to get started after 6 days try feeding it twice a day with each feed 12 hours apart.

Once active, store your starter in the fridge between uses – unless you bake every day or every other day – in which case it can be stored in a cool room.

## Sourdough Rye Bread

If you can't be bothered or haven't been able to build your own starter (see previous recipe) you'll need to get hold of a friend's mother (starter culture, I mean) or purchase some online. Either way, your starter will need to be active and ready to use; if not, wake it up and give it a couple of feeds 12 hours apart until it is nice and bubbly. To tell if your starter is active, drop a teaspoon of it into a bowl of cold water. It should float, but if it sinks, give it another feed and try again 8–10 hours later.

This recipe is adaptable to your own needs and conditions and works well with either white or wholemeal rye (you may need to use a little less water if using white). This version differs a little from that on page 141, allowing the loaf to rise directly in the baking tin so it only requires one transfer. Keep tweaking the timings and recipe to adapt to your taste and for your own needs and conditions – have fun with it!

MAKES 1 X 500G LOAF
*80g active sourdough starter*
*450g rye flour (light or wholemeal)*
*450–475ml tepid water*
*50g strong wholemeal or spelt flour*
*3 tsp sea salt*
*100g mixed seeds (sunflower, linseed, pumpkin and sesame)*
*Butter or olive oil, for greasing*

Start by making a pre-ferment. Spoon the active starter into a medium bowl with 100g of the rye flour and add 100ml tepid water. Mix well to combine, cover with a damp tea towel and set aside at room temperature overnight or for 12–24 hours – it should be bubbling and lively and small pinhole bubbles should have appeared on the top. If you have used a glass bowl you should see plenty of bubbles on the side of the mixture. While it's bubbling, put 3 tablespoons of the new mother back into the fridge.

In a large bowl combine the remaining 350g rye flour with the strong wholemeal flour and sea salt. Add the pre-ferment and 350–375ml tepid water and mix well to a spongy wet mass. Add a little more water if needed. Add 75g of the mixed seeds and mix to combine.

Butter or oil the inside of a 900g loaf tin (or cast-iron or heatproof casserole) and scatter with flour to stop it sticking. Scoop the dough into the tin and, using wet hands, press it level. Cover the tin with a clean tea towel or plastic shower cap (or pop inside a large plastic bag) and set aside at room temperature to rise for about 7–10 hours – or overnight in the fridge. The loaf should nearly have doubled in size and will have small pinhole bubbles on the top.

Preheat the oven to its highest setting – 240°C/220°C fan – and place a small roasting tin filled with water in the bottom of the oven. Spray the top of the loaf with water and sprinkle over the remaining seeds.

Bake the loaf on the middle shelf of the oven for 30 minutes and then reduce the temperature to 200°C/180°C fan and bake for a further 30 minutes until the loaf is deep golden brown.

Turn out onto a wire rack and leave to cool thoroughly before slicing. Ideally once cold wrap the loaf and leave overnight before cutting as the flavour will mellow and the texture will improve. Feel free to slice half and store in the freezer.

#### VARIATION: SUPER HEALTHY DIVERSITY LOAF

To add a greater diversity of plants and even more fibre and protein you can do what I now try to do regularly to improve my loaf by adding two extra ingredients to your flour. The first is to try swapping 25–50g of the regular flour with a special flour called Hodmedods Diversity XXX Blend, which contains ten plants and botanicals. This will give the loaf an extra level of texture and flavour complexity and will be a treat for your gut microbes. Another even simpler hack is to add 3–4 scoops (30–40g) of the ZOE prebiotic supplement called Daily 30+ that has over thirty-two dried plants (including nuts, seeds, herbs, spices and mushrooms) and adds a bit of colour to the loaf. Each scoop adds 5 grams of fibre. This is now my favourite loaf.

## Fermented Cashew Nut Cheese

An alternative way to ferment this cashew cheese is to add 1 tablespoon of your sauerkraut or kimchi brine or 1 tablespoon of white miso paste in place of the probiotic capsules.

MAKES 1 POT

*250g raw cashew nuts (not roasted or salted)*
*½ tsp sea salt*
*2 probiotic capsules (at least 4–5 billion* CFU*, containing lactic acid-forming microbes like* Lactobacilli*)*
*2–3 tsp nutritional yeast*
*2–3 tsp lemon juice*
*Freshly ground black pepper*
*½ tsp garlic granules*
*2 tsp dried or fresh chives*
*A pinch of dried chilli flakes (optional)*

Soak the cashew nuts in cold water overnight to soften them.

Strain the soaked nuts through a sieve, quickly rinse under cold running water and then tip into a food processor or blender. Add the salt and blend until smooth, scraping down the sides of the mixer from time to time.

Empty the contents of the probiotic capsules (or sauerkraut/kimchi brine or miso paste, if using) into the mixture, add the nutritional yeast and lemon juice and season with freshly ground black pepper. Blend again to combine.

Transfer to a Kilner or glass jar, making sure you leave about 3cm from the top of the cheese to the top of the jar. Cover and leave to ferment for 12–24 hours at room temperature.

*When ready the cheese should be fluffy with bubbles and smell slightly sour.*

Stir in the garlic, herbs and chilli flakes (if using), season to taste with salt and pepper and keep in the fridge. Use it like a soft cream cheese and enjoy within a week.

## Kenji's Miso (Red and White)

White miso is sweeter and has a less complex flavour than red miso; it can take as little as a week to ferment whereas the red miso will take months. I have included recipes for both red and white miso below. The method for each is the same but the quantities and fermenting times are different. Adding a little good-quality ready-made miso to the white miso mixture will help to kick-start fermentation. This is Kenji Morimoto's recipe.

---

### Red Miso

MAKES 500G

*150g dried soybeans (to yield 300g when cooked)*
*150g rice koji*
*108g sea salt (13 per cent final weight), plus 2 tsp*
*1 tbsp good miso paste*
*2–3 tbsp vodka (for cleaning the storage jar)*
*Extra uncooked regular rice, dried pulses or salt for weights (roughly 500g)*

---

### White Miso

MAKES 500G

*150g dried soybeans (to yield 300g when cooked)*
*300g rice koji*
*48g sea salt (8 per cent), plus 2 tsp*
*1 tbsp good miso paste (optional)*
*2–3 tbsp vodka (for cleaning the storage jar)*
*Extra uncooked regular rice, dried pulses or salt for weights (roughly 500g)*

Rinse the soybeans, tip into a large bowl, cover with fresh water and leave to soak for 12 hours or overnight.

Drain and rinse the beans and put into a saucepan. Cover with water and simmer for 4–6 hours, or until soft enough to be easily squashed between your fingers, topping up the water as necessary.

Drain the beans and reserve the cooking liquid. Leave both the beans and water to cool to room temperature.

Tip the beans into a food processor and blend into a smoothish paste. In a bowl mix the dried koji with the salt and tablespoon of miso paste and add to the beans. Blend again until combined, adding 2–3 tablespoons of the reserved cooking water if needed.

Using your clean hands shape the miso into tight balls – roughly the size of a satsuma – and pack them one at a time into a spotlessly clean jar. Use your knuckles to press the miso into the jar evenly to knock out any air pockets. Continue adding the balls to the jar and pack them into a smooth layer.

Clean the insides of the rim of the jar with some kitchen paper dampened with vodka or spirit and sprinkle 2 teaspoons of salt over the surface of the miso. Cover with a layer of cling film, pressing it onto the surface and neatly and tightly around the edges. Fill a plastic food bag with dried rice, beans or salt and place on top of the cling film to keep any air from the surface of the miso.

Cover with a lid and store in a cool spot away from light for 6 months to make red miso and for as little as 3 weeks to a few months for white miso.

Miso will keep for months if not years – keep in the fridge once optimum fermentation has been reached. I use it as a healthy stock cube, added to stews, soups, or on vegetables to enhance flavours such as sprouts or broccoli.

## Misozuke – Japanese Pickled Vegetables

Whether you buy it or make your own, it's great to see the power of unpasteurised miso used as a starter for fermenting vegetables such as purple sprouting broccoli, small broccoli or cauliflower florets, or kale. If you want to ferment more watery vegetables such as daikon or radishes, chop and dry or salt them first to draw out the excess moisture.

I have given relatively small quantities for the recipe below, but it can be scaled up to suit your needs. I use white miso paste, but any type of miso will work. The miso paste quantity should be 50 per cent of the weight of the prepared vegetables.

SERVES 2–3

*100g white miso paste*
*2 tbsp rice vinegar, mirin or sake*
*1 tbsp finely grated fresh ginger*
*1 garlic clove, crushed or finely grated*
*½ tsp dried chilli flakes*
*200g small broccoli florets*

Combine the miso and rice vinegar (or mirin or sake) in a bowl with the ginger, garlic and chilli flakes and mix to combine.

Add the broccoli and mix well to combine and thoroughly coat the vegetables with the paste.

Pack into a clean lidded jar or resealable plastic bag and place in the fridge. The pickles can be eaten in a few hours but are best anywhere between 4–7 days.

To serve, remove the vegetables from the paste and scrape off any excess. Serve as a side dish or hors d'oeuvre.

## Tim's Fermented Mushroom Pate

You could add some chopped soft herbs such as parsley, chives or tarragon to this pate, or even stir through a couple of tablespoons of ricotta or cream cheese once the fermented mushrooms have been blended.

MAKES 200–300G (SERVES 4)
*250g chestnut mushrooms, trimmed and roughly chopped*
*2–4 garlic cloves, chopped*
*4g sea salt*
*200ml filtered water*
*Salt and freshly ground black pepper*
*Dash of Tabasco (optional)*
*1–2 tbsp chopped soft herbs (optional)*
*1–2 tbsp ricotta or cream cheese (optional)*

Add the mushrooms into a dry frying pan and fry over a medium heat for about 4 minutes until browned and softening.

Tip the mushrooms into a clean fermenting jar, add the chopped garlic and leave to cool to room temperature. Meanwhile prepare the 2 per cent brine. Dissolve the salt in the filtered water and pour into the jar to cover the mushrooms. Add more brine if needed and weigh the mushrooms down to submerge in the brine.

Cover the jar and leave to ferment at room temperature but out of direct sunlight for 3–5 days, depending on the room temperature.

Taste after 3 days to check that the mushrooms are fermenting and then either continue to ferment if not or pour off most of the brine and save for another use. Whizz the mushrooms and garlic in a food processor until finely chopped and season to taste. Add a dash of Tabasco and some chopped herbs (if using) – or stir through a couple of tablespoons of ricotta or cream cheese to lighten the pate.

Serve spooned onto sourdough toast or crackers with a drizzle of extra virgin olive oil and a scattering of finely chopped herbs or other pickles. This pate is also delicious spread over my Sourdough Rye Bread (see page 252).

VARIATION: Skip the food processor stage and simply enjoy the mushrooms as pickles.

## Gherkins

Don't be tempted to use regular cucumbers here – they are too watery and you'll just end up with mush. You can make up your own mix of pickling spices or use just one spice; choose from mustard seeds, celery seeds, onion seeds, coriander seeds, cumin seeds, chilli flakes, black peppercorns, allspice, fennel seeds.

MAKES 1 MEDIUM JAR

*400g small/baby or pickling cucumbers*
*2 shallots, thinly sliced*
*1 garlic clove, thinly sliced*
*2–3 sprigs of fresh dill*
*1–2 bay leaves*
*About 25–30g sea salt*
*2 tsp pickling spices*

Wash the cucumbers, trim the ends and cut into halves or quarters down their length, depending on their size.

Place your jar on digital scales and set to zero. With the jar still on the scales arrange the shallots and garlic in the bottom and place the dill, bay leaves and pickling spices on top to hold them in place. Pack the cucumbers into the jar and pour in filtered water to cover the cucumbers by about 2cm. Make a note of the combined weight of vegetables and water.

Calculate 2 per cent of the total weight and add this amount of salt to the jar. Tightly cover the jar with a lid and give it a good shake to dissolve the salt.

Place a weight on top of the cucumbers and secure the lid again. Place the jar on a plate and leave at room temperature out of direct sunlight for 5–7 days, burping the jar as needed.

Taste the gherkins after 4–5 days – they should be crisp, tangy and sour – continue fermenting until the desired rate of fermentation has been reached (this could take up to 7 days). Remove the weight and store the gherkins in the fridge. Eat within 3–4 weeks.

## Lacto Chillies

For this recipe I would recommend using the larger regular chillies rather than Thai or bird's eye chillies. Dried chillies can be used or added to fresh ones after soaking for a few hours in brine. You can switch the spices around according to taste – try adding mustard or fennel seeds and bay leaves.

MAKES 1 LARGE JAR
*400g red and/or green chillies*
*3–4 garlic cloves*
*1 tsp black peppercorns*
*1 tsp cumin seeds*
*1 tsp coriander seeds*
*600ml filtered water*
*About 25g sea salt*

Tip the chillies into a colander and wash under running water. Leave to dry and then trim off the stalks, cut in half and remove the seeds.

Place a clean fermenting jar on your digital scales and set to zero. Pack the chillies and garlic cloves into the jar and pour in enough water to cover. Make a note of the total weight, then drain the chillies and garlic and return them to the dry jar along with the spices. Make up the brine using filtered water and sea salt – this should be 2–3 per cent of the total weight of water, chillies and garlic. For example, if the total weight is 900g you will need 22–23g salt.

Pour the brine over the chillies, top with a fermenting weight so that they are submerged, cover and ferment at room temperature and out of direct sunlight for 7–10 days. Taste after 7 days – the chillies should be tangy.

You can now either store the chillies in the fridge or strain off most of the brine and blitz the chillies and garlic until smooth to make a fermented chilli sauce. Store in the fridge and use within 3–4 months, either on their own as an appetizer or added to spice up other dishes if using as a paste.

## Lacto Tomatoes

Enjoy these flavour bombs in salads, stews or on toasted sourdough as a souped-up bruschetta or *pan con tomate.*

MAKES 1 LARGE JAR

*500g cherry tomatoes on the vine*
*2–3 garlic cloves*
*2–3 sprigs of herbs (rosemary, thyme, oregano, dill)*
*1 chilli, washed and stalk removed (optional)*
*About 20g sea salt*
*500ml filtered water*

Remove the tomatoes from the vine, wash them under running water and use a skewer to poke a few holes in each one.

Pack the tomatoes into a spotlessly clean glass jar with the whole peeled garlic cloves, herb sprigs and chilli (if using).

Make up a 4 per cent brine and pour into the jar to cover the tomatoes by 2cm. Place a weight on top of the tomatoes to keep them submerged and cover with a lid.

Place the jar on a plate and leave at room temperature but out of direct sunlight to ferment for 5–7 days. Taste a tomato after 5 days – if they are nice and fizzy and tangy then they can be moved to the fridge and eaten within 4 weeks.

## Other Lactoferments

The method for tomatoes can be applied to larger, green tomatoes, and to many other vegetables, such as green beans cut into quarters, or firmer plants like celeriac, carrots or fennel, shredded with a mandolin into thin slices. After weighing, pack them into a large jar and cover with 2cm of salted water (2 per cent of combined weight). Cover the vegetables with weights to submerge them below the water line, cover the jar with a slightly loose lid, and place it on a plate in a cool, dark cupboard.

Fermentation should occur in 7–10 days. The leftover brine is a very useful addition for other ferments or sauces.

## The Simplest Homemade Vinegar

This is an excellent way to use up a half-drunk bottle of wine, though you could also use beer or cider.

Decant the wine (red or white) into a clean wide-mouthed glass jar and dilute with roughly two parts filtered water to every three parts wine in order to get a final alcohol content of around 7 percent.

Cover with a clean cloth and secure with string or a rubber band. This will allow air to circulate in the jar which will allow microbes to do their work.

Leave at cool room temperature but out of direct sunlight for about 8 weeks, then decant into a narrow-necked bottle with a lid or cork stopper which will keep air out and stop further fermentation.

## Apple Scraps Cider Vinegar

This is a great easy way to avoid food waste and produce a great-tasting vinegar with potential health benefits.

MAKES 1 LITRE

*Peelings, cores and a little of the fruit from 3–5 apples, ideally organic*
*1 litre filtered water*
*100g sugar*

Put the apple scraps into a clean 1 litre jar. Make a 10 per cent sugar solution by stirring together the water and sugar, then pour this over the fruit. Submerge the fruit in the solution and cover the jar with a clean cloth secured with string or an elastic band.

Leave at room temperature but out of direct sunlight for 2–4 weeks. During this time the mixture will ferment and turn first into alcohol and then vinegar. The time will depend on the temperature of your room.

Strain the fruit and transfer the vinegar to a clean bottle with a cap or cork to keep air out. Store at room temperature and out of direct sunlight. Enjoy diluted as a drink or as a vinegar in salad dressings.

# Troubleshooting

Fermenting is part science, part art and a lot of trial and error as everyone's home environment is different. It is therefore vital that you remember what worked and what didn't. Having a fermenting notebook can be helpful, as are stickers on the bottles reminding you what you added and when. When researching fermenting recipes I often found quite considerable differences in some areas such as how much sugar to add to kombuchas and water kefirs, as well as timing, so feel free to experiment to your taste. I have so far not had any major problems, apart from some unwanted extra white yeasts and my ferments getting exposed to air more than I wanted. I've also had bad results when I've used vegetables that were past their best or bruised or damaged fruits.

Here are some tips I found useful. Remember that you should use your eyes and nose to tell you if something has gone wrong. If it looks bad and smells bad – it probably is!

## General fermenting

**Problem**: A thin white odourless film of froth covers your ferment, particularly krauts or fermented chillies.
**Solution**: It is likely a harmless kahm yeast floating around, so scrape it off and if needed rebottle; the rest is fine to eat.

**Problem**: A coloured mould forms on top of the ferment. Moulds are fuzzy and have an acrid smell with blue and green colours.
**Solution**: Although some people throw the top layer away and eat what is below the surface, I would chuck it out, start again and correct the problem. This is likely due to letting in too much oxygen,

not submerging the vegetables in brine properly, or sometimes being in a damp room with mould in the walls.

**Problem**: Ferment is too salty.
**Solution**: Maybe you added too much, or got your sums wrong. All is not lost. You can rinse it in water, add it to some unfermented cabbage or salad or use it for cooking.

**Problem**: I'm going away for a few months. Can I freeze my mothers and grains?
**Solution**: It depends. I have frozen my sourdough starter for a few months successfully. Water kefir doesn't freeze well as you lose some of the microbes that make the backbone of the grains. Milk kefir can be frozen but needs to be dried out first. A kombucha SCOBY lasts well in the back of the fridge for months.

## Yogurt

**Problem**: My yogurt failed to set properly.
**Solution**: This often happens when the milk was above the magic 46°C when the starter was added, or the incubation temperature dropped below 38°C so take more care or use a thermos or tea cosy to conserve heat.

**Problem**: My yogurt tastes fine but is runny.
**Solution**: This is probably because the temperature dropped too quickly and you need better insulation. Try using whole cow's milk instead heated to 82°C for 15 minutes to break up the proteins and evaporate some water.

**Problem**: My yogurt has curdled early on.
**Solution**: Some separation of the whey always occurs after a few days but if this happens early on it is likely due to keeping it at high temperatures for too long. Adjust your starting temperature next time.

## Milk kefir

**Problem**: Milk separates causing the kefir to form only in the top creamy part.
**Solution**: ensure milk is well shaken just before you add the kefir grains.

**Problem**: How do I know when it's ready?
**Solution**: This will depend on conditions, but the smell of sourness is a good guide as is seeing the grains float on top of the curd.

**Problem**: Pockets of lumpy whey have formed.
**Solution**: This can happen, especially in warm weather or if left too long. It's not a big problem and can be reincorporated with stirring or whisking.

**Problem**: Milk curdles into tiny curds.
**Solution**: This can happen in summer and the kefir loses its distinctive smell. It means other spoilage microbes have taken over and you need to abandon the kefir and retry in a cooler space.

**Problem**: My kefir is too sour.
**Solution**: Try fermenting for a shorter time and adding fewer grains to the milk next time.

**Problem**: How do I grow extra kefir grains to give to friends?
**Solution**: Add grains to a jar with 100ml of fresh milk and leave at room temperature for 3 days.

**Problem**: How long do grains last?
**Solution**: Indefinitely if you keep them well in a comfy 5-star hotel in your fridge. This should be in a jar with 10–20 times more milk than grains. Either use them as they are or refresh (and if enough separate them) with new milk every three weeks. I am

not a good hotelier and haven't managed for more than 6 months yet . . .

## Water kefir

**Problem:** My kefir looks thick and slimy in the glass and tastes sweet.
**Solution:** Probably because LAB microbes are making too much dextran because they are overfed with sugar or too warm. It will be safe to drink but cut back on sugar next time and add some lemon peel.

**Problem**: My kefir is still thin and clear after 3 days.
**Solution**: Suggests no fermentation and dead grains so try and resuscitate by adding grains to twice the volume of sugary 4 per cent water (with an optional fig) and leave for another 1–2 days. If this fails you can pronounce them dead.

**Problem**: My grains are going mushy.
**Solution**: Don't worry they still make decent kefir, but you can strain out the smaller ones, ideally with a plastic sieve as they don't like metal contact.

## Fermented vegetables

**Problem**: Can I open my sealed Kilner jar to see if fermentation is working?
**Solution**: Yes you can, but wait for at least 72 hours (longer for chillies) otherwise you let oxygen in and could mess things up and attract kahm yeast.

**Problem**: My top layer of vegetables has an odd colour.
**Solution**: They were probably not submerged properly – a common problem. Next time use a plastic lid, ziplock bag filled with salty water, sterilised stones or commercial weights.

**Problem**: There's a sulphur smell when opening or burping the jar or container that dissipates.
**Solution**: Not a real problem, as vegetables like cabbage produce gases like sulphur. Get used to it!

**Problem**: Continuous strange rotten smell that hangs around.
**Solution**: Oh dear. Maybe you underdid the salt and rot has set in, or maybe try organic veg next time.

**Problem**: The vegetables are going slimy.
**Solution**: Throw them away. Usually a sign of early rot with not enough salt.

## Kombucha

**Problem**: My SCOBY looks mottled and ugly.
**Solution**: A healthy blob can have light and dark brown rings and tendrils hanging off the bottom. If it has moulds on the top, it is either a sign it wasn't kept in enough liquid or it has got sick and needs to be retired.

**Problem**: My SCOBY sank when I put in the liquid – is it dead?
**Solution**: No, I worried the first time, but it normally rights itself and rises after a few days.

**Problem**: It smells funny . . . is it supposed to have farmyard and nail polish notes?
**Solution**: No and it shouldn't smell of vomit either. It clearly did not produce acid quickly and got contaminated. Throw it and the infected SCOBY away and grow another from your hotel. Adding some mature kombucha to the new mix helps protect it from this problem.

**Problem**: My kombucha tastes ok but is too flat.
**Solution**: Secondary fermentation (as I discovered) is the key to fizz. Fill your bottles and stir to shake the yeast around with a small

teaspoon of sugar (plus optional fruit) and leave at room temperature for 1–3 days before refrigeration. Note this is easily overdone, especially if using peach or mango as I discovered yesterday to my cost when I lost half a bottle.

**Problem**: It tastes too much like vinegar.
**Solution**: This means it probably is vinegar and you left it too long. Try using it as a vinegar by adding a teaspoon of sugar with the SCOBY still floating and taste again in 2 weeks. Or try mixing it with your water kefir to spice it up.

**Problem**: My SCOBY developed a thin film across the top.
**Solution**: This is probably a baby SCOBY you can leave or peel off and add to your hotel or give to a friend.

## Sourdough

**Problem**: My mother smells a bit vinegary after being in the fridge and has some liquid in it.
**Solution**: Don't worry this is common and every mother is different. Wake it up with fresh flour and water and it should be ok. Only worry if it has other strange smells or changes to a very dark colour.

**Problem**: My bread doesn't rise like it should, and I can't get the big pockets of air.
**Solution**: Don't worry. I often couldn't at first, but it still tasted good. There are a few options to try: leave the mother longer until it's bubbling and really active; try folding the dough a few times but don't knead it; rest the bread longer in the greased oven dish before cooking. Changing to better organic flour can help too. But basically it's trial and error and can just improve as the mother (and you) matures.

**Problem**: My crust is too tough.
**Solution**: Buy a better bread knife or try wrapping the baked loaf in a damp tea towel just after baking.

**Problem**: My rye bread is still too moist after cooking.
**Solution**: Try toasting it! Or leave it longer cooking with the lid off, and let it cool slowly in the oven with the door open for 10 minutes.

**Problem**: My bread gets stuck to the baking dish and I damage it on removing.
**Solution**: Coat or spray the dish well with a vegetable oil and sprinkle over some flour.

# Further reading and resources

These are some of the books I recommend if you want to try more fermenting recipes:


Cooper, Sam, *The Fermentation Kitchen* (DK, 2024)

Gilmartin, Caroline, *Fermented Foods: A Practical Guide* (The Crowood Press, 2020)

Katz, Sandor, *The Art of Fermentation* (Chelsea Green Publishing, 2012)

Kimbell, Vanessa, *10-Minute Sourdough* (Kyle Books, 2021)

Morimoto, Kenji, *Ferment: Simple Ferments and Pickles and How to Eat Them* (Pan Macmillan, 2025)

Parker, Andy, *CAMRA's Essential Home Brewing* (CAMRA Books, 2018)

Read, James, *Of Cabbages & Kimchi: A Practical Guide to the World of Fermented Food* (Particular Books, 2023)

Simonsson, Asa, *Fermentation* (Lorenz Books, 2019)

Skinner, Julia, *Our Fermented Lives* (Storey Publishing, 2022)

de Thample, Rachel, *Fermentation: River Cottage Handbook No. 18* (Bloomsbury Publishing, 2020)

Wise, Jaega, *Wild brews: The Craft of Home Brewing* (Kyle Books, 2022)

# Notes

## PART ONE

7 **Called ultramicrobiota (UMB)**: H.-W. Lee, 'Presence of an ultra-small microbiome in fermented cabbages', *PeerJ* (2023); 17:11

8 **newly discovered 'neuropod' cells**: M. M. Kaelberer, 'Neuropod cells: The emerging biology of gut-brain sensory transduction', *Annual Review of Neuroscience* (2020); 43

10 **they can improve health**: S. M. Gibbons, 'Perspective: Leveraging the gut microbiota to predict personalized responses to dietary, prebiotic, and probiotic interventions', *Advances in Nutrition* (2022); 13(5)

12 **shrunken versions of themselves**: H. Shi, 'Starvation induces shrinkage of the bacterial cytoplasm', *Proceedings of the National Academy of Sciences of the United States of America*, (2021); 118(24)

13 **in gut samples of 30,000 people**: N. Carlino, 'Unexplored microbial diversity from 2,500 food metagenomes and links with the human microbiome', *Cell* (2024); 187(20)

15 **health benefits of seaweed**: N. A. Pudlo, 'Diverse events have transferred genes for edible seaweed digestion from marine to human gut bacteria', *Cell Host Microbe* (2022); 30(3)

15 **The microbe *Lawsonibacter* is found in coffee**: P. Manghi, 'Coffee consumption is associated with intestinal *Lawsonibacter asaccharolyticus* abundance and prevalence across multiple cohorts', *Nature Microbiology* (2024); 9

20 **adults have inside them**: I. Mogilnicka, 'Gut mycobiota and fungal metabolites in human homeostasis', *Current Drug Targets* (2019); 20(2)

20 **failing a breathalyser test**: R. J. Dinis-Oliveira, 'The auto-brewery syndrome: A perfect metabolic "storm" with clinical and forensic implications', *Journal of Clinical Medicine* (2021); 10(20)

22 **Viruses are everywhere in our food**: B. L. Maske, 'Viruses in fermented foods: are they good or bad? Two sides of the same coin', *Food Microbiology* (2021); 98

22 **eating 100 billion phages**: Q. Wu, 'Phages in fermented foods: interactions and applications', *Fermentation* (2023); 9:201

25 **after much heated discussion**: S. Salminen, 'The International Scientific Association of Probiotics and Prebiotics (ISAPP) consensus statement on the definition and scope of postbiotics', *Nature Reviews Gastroenterology & Hepatology* (2021); 18

## PART TWO

31 **study of different primate species**: K. L. Bryant, 'Fermentation technology as a driver of human brain expansion', *Communications Biology* (2023); 6

33 **can carry nasty strains**: K. Skowron, 'Two faces of fermented foods – the benefits and threats of its consumption', *Frontiers in Microbiology* (2022); 13

34 **a few isolated outbreaks**: S.-O. Kim, 'Recent (2011–2017) foodborne outbreak cases in the Republic of Korea compared to the United States: a review', *Food Science and Biotechnology* (2021); 30(2)

34 **cheese reaction was linked**: M. Alkhouli, 'Revisiting the "cheese reaction": more than just a hypertensive crisis?', *Journal of Clinical Psychopharmacology* (2014); 34(5)

34 **headaches falsely claimed**: V. T. Martin, 'Diet and headache: Part 1', *Headache* (2016); 56(9)

34 **linked to their microbiome**: K. G. Jameson. 'Toward understanding microbiome-neuronal signaling', *Molecular Cell* (2020); 78(4)

35 **a Japanese commercial study**: Y. Kotani. 'Oral intake of *Lactobacillus pentosus* strain b240 accelerates salivary immunoglobulin A secretion in the elderly: A randomized, placebo-controlled, double-blind trial', *Immunity & Ageing* (2010); 7(11)

35 **in common cold infections**: S. Shinkai, 'Immunoprotective effects of oral intake of heat-killed *Lactobacillus pentosus* strain b240 in elderly adults: a randomised, double-blind, placebo-controlled trial', *British Journal of Nutrition* (2013); 109(10)

35 **finding was only in mice**: W. S. Hong, 'Effect of heat-inactivated kefir-isolated *Lactobacillus kefiranofaciens* M1 on preventing an allergic airway response in mice', *Journal of Agricultural and Food Chemistry* (2011); 59(16)

35 **summarised seven randomised controlled**: J. N. Malagón-Rojas, 'Postbiotics for preventing and treating common infectious diseases in children: A systematic review', *Nutrients* (2020); 12(2)

36 **appeared to improve gut health**: P. Roggero, 'Analysis of immune, microbiota and metabolome maturation in infants in a clinical trial of *Lactobacillus paracasei* CBA L74-fermented formula', *Nature Communications* (2020); 11

36 **reduced symptoms after eight weeks**: V. Andresen, 'Heat-inactivated *Bifidobacterium bifidum* MIMBb75 (SYN-HI-001) in the treatment of irritable bowel syndrome: a multicentre, randomised, double-blind, placebo-controlled clinical trial', *Lancet Gastroenterology Hepatology* (2020); 5(7)

36 **these were not freak results**: F. Balaguer, 'Lipoteichoic acid from *Bifidobacterium animalis* subsp. *lactis* BPL1: A novel postbiotic that reduces fat deposition via IGF-1 pathway', *Microbial Biotechnology* (2022); 15(3)

36 **study found that it**: Y. Zhong, 'The protective effect of heat-inactivated *Companilactobacillus crustorum* on dextran sulfate sodium-induced ulcerative colitis in mice', *Nutrients* (2023); 15:2746

36 **when a commercial milk kefir**: B. Bourrie, 'Consumption of kefir made with traditional microorganisms resulted in greater improvements in LDL cholesterol and plasma markers of inflammation in males when compared to a commercial kefir: A randomized pilot study', *Applied Physiology, Nutrition, and Metabolism* (2023); 48

36 **has some antimicrobial activity**: A. R. Al-Mohammadi, 'Chemical constitution and antimicrobial activity of kombucha fermented beverage', *Molecules* (2021); 26(16)

36 **for irritable bowel syndrome (IBS) sufferers**: N. E. Nielsen, 'Lacto-fermented sauerkraut improves symptoms in IBS patients independent of product pasteurisation – a pilot study', *Food & Function* (2018); 9(10)

36 **twin studies in 2014**: J. K. Goodrich, 'Human genetics shape the gut microbiome', *Cell* (2014); 159(4)

37 **greater than when it was alive**: P. D. Cani, 'Akkermansia muciniphila: paradigm for next-generation beneficial microorganisms', *Nature Reviews Gastroenterology & Hepatology* (2022); 19(10)

37 **stimulates the anti-hunger gut hormone**: H. S. Yoon, 'Akkermansia muciniphila secretes a glucagon-like peptide-1-inducing protein that improves glucose homeostasis and ameliorates metabolic disease in mice', *Nature Microbiology* (2021); 6(5)

37 **integrity of the gut barrier**: H. Wade, 'Akkermansia muciniphila and its membrane protein ameliorates intestinal inflammatory stress and promotes epithelial wound healing via CREBH and miR-143/145', *Journal of Biomedical Science* (2023); 30(1)

37 **alter energy and blood sugar control**: C. Depommier, 'Beneficial effects of *Akkermansia muciniphila* are not associated with major changes in the circulating endocannabinoidome but linked to higher mono-palmitoyl-glycerol levels as new PPARα agonists', *Cells* (2021); 10(1)

37 **by reducing inflammation in the blood**: D. E. Motei, 'Supplementation with postbiotic from *Bifidobacterium breve* BB091109 improves inflammatory

status and endocrine function in healthy females: a randomized, double-blind, placebo-controlled, parallel-groups study', *Frontiers in Microbiology* (2023); 14

37 **and immune dendritic cells**: D. Finnegan, 'Novel dairy fermentates have differential effects on key immune responses associated with viral immunity and inflammation in dendritic cells', *Foods* (2024); 13

38 **similar to statins**: S. Makino, 'Enhanced natural killer cell activation by exopolysaccharides derived from yogurt fermented with *Lactobacillus delbrueckii* ssp. *bulgaricus* OLL1073R-1', *Journal of Dairy Science* (2023); 99(2)

38 **replicated in humans**: S. J. Park, 'Postbiotics against obesity: perception and overview based on pre-clinical and clinical studies', *International Journal of Molecular Sciences* (2023); 24(7)

38 **useful as anti-cancer agents**: J. Wu, 'The anti-cancer effects and mechanisms of lactic acid bacteria exopolysaccharides *in vitro*: A review', *Carbohydrate Polymers* (2021); 253

41 **for heart and metabolic health**: F. Asnicar, 'Gut microbes linked with metabolic health, nutrition and diet intervention', *Nature* (2025)

41 **even more powerful in terms of gut benefits**: S. A. Ibrahim, 'A review and comparative perspective on health benefits of probiotic and fermented foods', *International Journal of Food Science and Technology* (2023); 58(10)

43 **taking probiotics or eating fermented foods**: P. Louca, 'Modest effects of dietary supplements during the COVID-19 pandemic: insights from 445,850 users of the COVID-19 Symptom Study app', *BMJ Nutrition, Prevention & Health* (2021); 4(1)

43 **following a high-fibre diet**: H. C. Wastyk, 'Gut-microbiota-targeted diets modulate human immune status', *Cell* (2021); 184(16)

43 **future anti-viral therapies**: K. Tillisch, 'Consumption of Fermented Milk Product With Probiotic Modulates Brain Activity', *Gastroenterology* (2013); 144(7)

44 **suggesting that fermented foods**: A. K. Paul; 'Are Fermented Foods Effective against Inflammatory Diseases?', *International Journal of Environmental Research and Public Health* (2023); 20(3)

45 **when the tumour has started**: S. Ugai, 'Long-term yogurt intake and colorectal cancer incidence subclassified by *Bifidobacterium* abundance in tumor', *Gut Microbes* (2025); 17(1)

45 **after drinking fermented milks**: K. Tillisch, 'Consumption of fermented milk product with probiotic modulates brain activity', *Gastroenterology* (2013); 144(7)

45 **reduced perceived stress in twenty-four**: K. Berding, 'Feed your microbes to deal with stress: A psychobiotic diet impacts microbial stability

and perceived stress in a healthy adult population', *Molecular Psychiatry* (2023); 28(2)

48 **actually very healthy for our immune system**: T. Newman, 'Why are short-chain fatty acids important', ZOE (2024)

49 **LAB microbes can often produce GABA**: B. D. Palanisamy, 'Gamma-aminobutyric acid (GABA) production by potential probiotic strains of indigenous fermented foods origin and RSM based production optimization', *LWT* (2023); 176

49 **been associated with mood**: Y. S. Jie, 'Antidepressive mechanisms of probiotics and their therapeutic potential', *Frontiers in Neuroscience* (2020); 13

49 **how many antidepressant medications**: J. Moncrieff; 'The serotonin theory of depression: a systematic umbrella review of the evidence', *Molecular Psychiatry* (2023); 28

50 **marker of inflammation and high lipids**: P. Louca, 'The secondary bile acid isoursodeoxycholate correlates with post-prandial lipemia, inflammation, and appetite and changes post-bariatric surgery', *Cell Reports Medicine* (2023); 4(4)

50 **influenced by these bile acids**: P. Gao, 'Gut microbial metabolism of bile acids modifies the effect of Mediterranean diet interventions on cardiometabolic risk in a randomized controlled trial', *Gut Microbes* (2024); 16(1)

50 **other common fermenting microbes**: Z. Huang, 'Unveiling and harnessing the human gut microbiome in the rising burden of non-communicable diseases during urbanization', *Gut Microbes* (2023); 15(1)

51 **chemicals then interact with gut lining**: A. Nakamura, 'Symbiotic polyamine metabolism regulates epithelial proliferation and macrophage differentiation in the colon', *Nature Communications* (2021); 12

51 **levels decrease with age**: L. Yu, 'Gut microbiota and anti-aging: Focusing on spermidine', *Critical Reviews in Food Science and Nutrition* (2024); 64(28)

51 **combinations of fermented foods**: Y. Kitada, 'Bioactive polyamine production by a novel hybrid system comprising multiple indigenous gut bacterial strategies', *Science Advances* (2018); 4(6)

52 **anti-inflammatory and antioxidant effects**: S. De Marco, 'Probiotic cell-free supernatants exhibited anti-inflammatory and antioxidant activity on human gut epithelial cells and macrophages stimulated with LPS', *Evidence-Based Complementary and Alternative Medicine* (2018)

52 **could inhibit colon cancer**: J. Escamilla, 'Cell-free supernatants from probiotic *Lactobacillus casei* and *Lactobacillus rhamnosus* GG decrease colon cancer cell invasion in vitro', *Nutrition and Cancer* (2012); 64(6)

54 **so-called niche theory explains**: R. R. Segura Munoz, 'Experimental evaluation of ecological principles to understand and modulate the

outcome of bacterial strain competition in gut microbiomes', *The ISME Journal* (2022); 16

## PART THREE

63 **our earliest foray into probiotic therapy**: A. C. Brown, 'Probiotics and medical nutrition therapy', *Nutrition in Clinical Care* (2004); 7(2)

64 **Cultured microbes in yogurt didn't**: N. Zmora, 'Personalized gut mucosal colonization resistance to empiric probiotics is associated with unique host and microbiome features', *Cell* (2018); 174(6)

64 **lower risk of cardiovascular disease**: L. Wu, D. Sun, 'Consumption of yogurt and the incident risk of cardiovascular disease: A meta-analysis of nine cohort studies', *Nutrients* (2017); 9(3)

64 **no clear benefits found for cancer**: H. Tutunchi, 'Yogurt consumption and risk of mortality from all causes, CVD and cancer: a comprehensive systematic review and dose-response meta-analysis of cohort studies', *Public Health Nutrition* (2023); 26(6)

64 **may also improve levels of blood sugar**: A. A. Dumas, 'A systematic review of the effect of yogurt consumption on chronic diseases risk markers in adults', *European Journal of Nutrition* (2017); 56(4)

65 **eating yogurt showed weight loss**: C. Sayon-Orea, 'Associations between yogurt consumption and weight gain and risk of obesity and metabolic syndrome: A systematic review', *Advances in Nutrition* (2017); 8(1)

65 **yogurt-eating habits of nearly 2,000 UK twins**: C. I. Le Roy, 'Yoghurt consumption is associated with transient changes in the composition of the human gut microbiome', (2020), PREPRINT (Version 1) available at www.researchsquare.com/article/rs-38248/v1

65 **protective against developing internal belly fat**: J. Zierer, 'The fecal metabolome as a functional readout of the gut microbiome', *Nature Genetics* (2018); 50(6)

65 **significantly reduced their abdominal fat**: T. Toshimitsu, 'Ingesting yogurt containing *Lactobacillus plantarum* OLL2712 reduces abdominal fat accumulation and chronic inflammation in overweight adults in a randomized placebo-controlled trial', *Current Developments in Nutrition* (2021); 5(2)

65 **fight a range of respiratory and gut viruses**: K. P. Kanmani, 'Exopolysaccharides from *Lactobacillus delbrueckii* OLL1073R-1 modulate innate antiviral immune response in porcine intestinal epithelial cells', *Molecular Immunology* (2018); 93

65 **improved markers of inflammation and immunity**: R. Pei, 'Low-fat yogurt consumption reduces biomarkers of chronic inflammation and inhibits markers of endotoxin exposure in healthy premenopausal women: a

randomised controlled trial', *British Journal of Nutrition* (2017); 118(12); R. Pei, 'Premeal low-fat yogurt consumption reduces postprandial inflammation and markers of endotoxin exposure in healthy premenopausal women in a randomized controlled trial', *Journal of Nutrition* (2018); 148(6)

65 **people taking probiotics were**: L. Panayiotis, 'Modest effects of dietary supplements during the COVID-19 pandemic: insights from 445,850 users of the COVID-19 Symptom Study app', *BMJ Nutrition, Prevention & Health* (2021)

66 **better response to the flu vaccine**: J. Hemmi, 'Consumption of yogurt fermented with *Lactobacillus delbrueckii* ssp. *bulgaricus* OLL1073R-1 augments serum antibody titers against seasonal influenza vaccine in healthy adults', *Bioscience of Microbiota, Food and Health* (2023); 42(1)

66 **gut-friendly and plant-diverse diet**: S. Yan, 'Characterization of the composition variation of healthy human gut microbiome in correlation with antibiotic usage and yogurt consumption', *Antibiotics (Basel)* (2022); 11(12)

66 **even higher for children's yogurts**: J. B. Moore, 'High levels of sugar in organic and children's yogurts – new survey', The Conversation (2018)

66 **the sugar these products contain**: J. B. Moore, 'Evaluation of the nutrient content of yogurts: a comprehensive survey of yogurt products in the major UK supermarkets', *BMJ Open* (2018); 8(8)

68 **US study examined 249 non-dairy yogurt**: W. J. Craig, 'Nutritional content and health profile of non-dairy plant-based yogurt alternatives', *Nutrients* (2021); 13

69 **microbial composition of plant-based yogurts**: E. S. Lim, 'Preparation and functional properties of probiotic and oat-based synbiotic yogurts fermented with lactic acid bacteria', *Applied Biological Chemistry* (2018); 61

69 **microbes might not survive as well**: M. M. Ayyash, 'Characterization of new probiotics from dairy and nondairy products—Insights into acid tolerance, bile metabolism and tolerance, and adhesion capability', *Journal of Dairy Science* (2021); 104(8)

73 **to identify the microbes in traditional milk kefir**: A. M. Walsh, 'Microbial succession and flavor production in the fermented dairy beverage kefir', *mSystems* (2016); 1(5)

73 **studying these fascinating grains**: B. Richard, 'A whole new generation of milk kefir grains formed from freeze-dried starter cultures – a fascinating insight into a hidden world', Research report (2016)

75 **many positive kefir studies**: D. D. Rosa, 'Milk kefir: nutritional, microbiological and health benefits', *Nutrition Research Reviews* (2017); 30(1)

75 **improvements to serotonin levels**: H.-L. Chen, 'Kefir peptides exhibit antidepressant-like activity in mice through the BDNF/TrkB pathway', *Journal of Dairy Science* (2021); 104(6)

75 **on lowering levels of blood glucose**: B. C. T. Bourrie, 'Kefir in the prevention and treatment of obesity and metabolic disorders', *Current Nutrition Reports* (2020); 9; D. A. Savaiano, 'Yogurt, cultured fermented milk, and health: a systematic review', *Nutrition Reviews* (2021); 79(5)

75 **benefits of kefir on protecting the gut lining**: Z. J. Pražnikar, 'Effects of kefir or milk supplementation on zonulin in overweight subjects', *Journal of Dairy Science* (2020); 103(5)

75 **significant reductions in blood inflammation**: H. C. Wastyk, 'Gut-microbiota-targeted diets modulate human immune status', *Cell* (2021); 184(16)

78 **stronger potential action against other infectious microbes**: H. Demir, 'Comparison of traditional and commercial kefir microorganism compositions and inhibitory effects on certain pathogens', *International Journal of Food Properties* (2020); 23(1)

78 **when fed kefir grains that were heat-treated**: B. C. T. Bourrie, 'Consumption of the cell-free or heat-treated fractions of a pitched kefir confers some but not all positive impacts of the corresponding whole kefir', *Frontiers in Microbiology* (2022); 13

79 **generally do not contain**: A. M. Walsh, 'Microbial succession and flavor production in the fermented dairy beverage kefir', *mSystems* (2016); 1

79 **and lower potential cardiac risk**: B. C. T. Bourrie, 'Consumption of kefir made with traditional microorganisms resulted in greater improvements in LDL cholesterol and plasma markers of inflammation in males when compared to a commercial kefir: a randomized pilot study', *Applied Physiology, Nutrition, and Metabolism* (2023); 48(9)

80 **more likely to put you in hospital**: S. Costard, 'Outbreak-related disease burden associated with consumption of unpasteurized cow's milk and cheese, United States, 2009–2014', *Emerging Infectious Diseases* (2017); 23(6)

86 **commonly found in all samples**: O. Yerlikaya, 'The metagenomic composition of water kefir microbiota', *International Journal of Gastronomy and Food Science* (2022); 30

87 **general improvement in the gut barrier**: M. Calatayud, 'Water kefir and derived pasteurized beverages modulate gut microbiota, intestinal permeability and cytokine production in vitro', *Nutrients* (2021); 13(11)

92 **more microbes actually reside on the garlic**: S. B. Lim, 'Garlic is a source of major lactic acid bacteria for early-stage fermentation of cabbage-kimchi', *Food Science and Biotechnology* (2015); 24

93 **and hundreds of different species**: E. Kim, 'Analysis of cultivable microbial community during kimchi fermentation using MALDI-TOF MS', *Foods* (2021); 10(5)

93 **and potentially many more useful compounds**: S. H. Lee, 'Unraveling microbial fermentation features in kimchi: from classical to meta-omics approaches', *Applied Microbiology and Biotechnology* (2020); 104

94 **Korean chilli has its own identity**: James Read, *Of Cabbages & Kimchi* (Particular Books, 2023)

95 **the highest fermented cabbage eaters**: Jean Bousquet, 'Cabbage and fermented vegetables: From death rate heterogeneity in countries to candidates for mitigation strategies of severe COVID-19', *Allergy* (2021); 76

95 **part of the original kraut mix**: E. S. Nielsen, 'Lacto-fermented sauerkraut improves symptoms in IBS patients independent of product pasteurisation – a pilot study', *Food & Function* (2018); 9(10)

96 **gave ninety people three types of kimchi**: H. Y. Kim, 'Kimchi improves irritable bowel syndrome: results of a randomized, double-blind placebo-controlled study', *Food & Nutrition Research* (2022); 66

96 **clinical study is worth mentioning**: S. Lim, 'Effect of *Lactobacillus sakei*, a probiotic derived from kimchi, on body fat in Koreans with obesity: A randomized controlled study', *Endocrinology and Metabolism* (2020); 35(2)

97 **following a Mediterranean diet**: C. Haro, 'Two healthy diets modulate gut microbial community improving insulin sensitivity in a human obese population', *Journal of Clinical Endocrinology & Metabolism* (2016); 101(1)

98 **kimchi scares are a result of the massive consumption**: S. O. Kim, 'Recent (2011–2017) foodborne outbreak cases in the Republic of Korea compared to the United States: a review', *Food Science and Biotechnology* (2021); 30(2)

98 **total salt intakes also fell**: S. Y. Kim, 'Nineteen-year trends in fermented food consumption and sodium intake from fermented foods for Korean adults from 1998 to 2016', *Public Health Nutrition* (2020); 23(3)

99 **no association between blood pressure**: H. J. Song, 'Consumption of kimchi, a salt fermented vegetable, is not associated with hypertension prevalence', *Journal of Ethnic Foods* (2014); 1(1)

99 **compared to salt reduction alone**: X. Yin, 'Effects of salt substitutes on clinical outcomes: a systematic review and meta-analysis', *Heart* (2022); 10

99 **reduce blood pressure by a small**: A. M. Lewis-Mikhael, 'Effect of Lactobacillus plantarum containing probiotics on blood pressure: A systematic review and meta-analysis', *Pharmacological Research* (2020); 153

105 **volunteers preferred the raw milk**: P. Rashtchi, 'Comparison of the microbial, physicochemical, and sensorial properties of raw and pasteurized Lighvan cheeses during ripening time', *Food Science & Nutrition* (2021); 9(10)

105 **clear differences in taste profiles**: N. I. P. Valente, 'Cheeses made from raw and pasteurized cow's milk analysed by an electronic nose and an electronic tongue', *Sensors (Basel)* (2018); 18(8)

107 **the main culprit in both being listeriosis**: M. Sebastianski, 'Disease outbreaks linked to pasteurized and unpasteurized dairy products in Canada and the United States: a systematic review', *Canadian Journal of Public Health* (2022); 113(4)

108 **slight reductions in heart disease**: A. Giosuè, 'Consumption of dairy foods and cardiovascular disease: a systematic review', *Nutrients* (2022); 14(4)

109 **symptoms in sensitive individuals**: Y. O. Shulpekova, 'Food intolerance: The role of histamine', *Nutrients* (2021); 13(9)

109 **it's also higher in sheep compared**: M. C. Ferrante, 'Focus on histamine production during cheese manufacture and processing: a review', *Food Chemistry* (2023); 419

111 **until he was arrested**: N. Squires, 'Italy's "Mozzarella King" arrested over "contaminated cheese"', *Telegraph* (2012)

111 **mozzarellas contained foreign cow's milk**: J. Haworth, 'One in four Italian mozzarella cheeses contain foreign milk products', *Mirror* (2016)

111 **selling analogue cheese as real mozzarella**: B. Vonow, 'FAKING IT: 'How takeaways are selling you pizzas topped with FAKE CHEESE', *Sun* (2016)

112 **consumers often prefer vegan cheese**: Good Food Team, 'The best vegan "cheese" taste tested 2023', BBC Good Food (2023)

115 **two to three times more volatile flavour**: A. R. Al-Mohammadi, 'Chemical constitution and antimicrobial activity of Kombucha fermented beverage', *Molecules* (2021); 26(16)

115 **gene sequencing of several samples**: M. Arıkan, 'Microbial composition of Kombucha determined using amplicon sequencing and shotgun metagenomics', *Journal of Food Science* (2020); 85

115 **found rich microbe diversity**: J. Yang, 'Microbial and chemical profiles of commercial Kombucha products', *Nutrients* (2022); 14(3)

115 **as well as purine alkaloids**: T. Esatbeyoglu, 'Additional advances related to the health benefits associated with kombucha consumption', *Critical Reviews in Food Science and Nutrition* (2023); 64(18)

117 **even dead microbes can be helpful**: A. R. Al-Mohammadi, 'Chemical constitution and antimicrobial activity of Kombucha fermented beverage', *Molecules* (2021); 26(16)

118 **highly publicised review of kombucha**: E. Ernst, 'Kombucha: a systematic review of the clinical evidence', *Forsch Komplementarmed Klass Naturheilkd* (2003); 10(2)

119 **kombucha drunk in normal amounts**: J. M. Kapp, 'Kombucha: a systematic review of the empirical evidence of human health benefit', *Annals of Epidemiology* (2019); 30

119 **reduction in blood sugar and insulin levels**: F. S. Atkinson, 'Glycemic index and insulin index after a standard carbohydrate meal consumed with live kombucha: A randomised, placebo-controlled, crossover trial', *Frontiers in Nutrition* (2023); 10

119 **ability to ward off unwanted microbes**: L. T. Phung, 'Changes in the chemical compositions and biological properties of kombucha beverages made from black teas and pineapple peels and cores', *Scientific Reports* (2023); 13(1)

126 **due to the compounds they produce**: E. Aykın, 'Bioactive components of mother vinegar', *Journal of the American College of Nutrition* (2015); 34(1)

126 **drinking vinegar daily**: A. Samad, 'Therapeutic effects of vinegar: a review', *Current Opinion in Food Science* (2016); 8

126 **have fed vinegar to chickens**: P. Allahdo, 'Effect of probiotic and vinegar on growth performance, meat yields, immune responses, and small intestine morphology of broiler chickens', *Italian Journal of Animal Science* (2018); 17(3)

126 **vinegar had the same effects**: H. O. Santos, 'Vinegar (acetic acid) intake on glucose metabolism: a narrative review', *Clinical Nutrition ESPEN* (2019); 32

127 **GABA levels are further increased**: P. B. Devi, 'Gamma-aminobutyric acid (GABA) production by potential probiotic strains of indigenous fermented foods origin and RSM based production optimization', *LWT* (2023); 176

127 **showing that GABA was important**: M. Nishimura, 'Effects of white rice containing enriched gamma-aminobutyric acid on blood pressure', *Journal of Traditional and Complementary Medicine* (2016); 6(1)

127 **with high levels of the bioactive compound**: S. Sugiyama, 'Hypotensive effect and safety of brown rice vinegar with high concentration of GABA on mild hypertensive subjects', *Japanese Pharmacology and Therapeutics* (2008); 36

127 **related to the combination of vinegar and bonito**: H. Tanaka, 'The effects of γ-aminobutyric acid, vinegar, and dried bonito on blood pressure in normotensive and mildly or moderately hypertensive volunteers', *Journal of Clinical Biochemistry and Nutrition* (2009); 45

127 **did not find significant reductions**: T. Kondo, 'Vinegar intake reduces body weight, body fat mass, and serum triglyceride levels in obese Japanese subjects', *Bioscience, Biotechnology, and Biochemistry* (2009); 73(8)

127 **showed a consistent effect on lowering**: A. Hadi, 'The effect of apple cider vinegar on lipid profiles and glycemic parameters: a systematic review and meta-analysis of randomized clinical trials', *BMC Complementary Medicine and Therapies* (2021); 21(1)

127 **patients with type 2 diabetes**: P. Mitrou, 'Vinegar consumption increases insulin-stimulated glucose uptake by the forearm muscle in humans with type 2 diabetes', *Journal of Diabetes Research* (2015)

129 **suggesting that the acidity of the vinegar**: H. O. Santos, 'Vinegar (acetic acid) intake on glucose metabolism: A narrative review', *Clinical Nutrition ESPEN* (2019); 32

130 **can lead to oesophageal ulcers**: J. Chang, 'Corrosive esophageal injury due to a commercial vinegar beverage in an adolescent', *Clinical Endoscopy* (2020); 53(3)

130 **taking some blood pressure drugs**: A. M. Rana, 'Severe metabolic acidosis: A case of triple hit with ketogenic diet, vinegar, and metformin in an obese patient', *Case Reports in Nephrology* (2020)

130 **more polyphenols and antioxidant compounds**: E. Aykın, 'Bioactive components of mother vinegar', *Journal of the American College of Nutrition* (2015); 34(1)

132 **study looked at the process of tepache fermentation**: W. Gutiérrez-Sarmiento, 'Microbial community structure, physicochemical characteristics and predictive functionalities of the Mexican tepache fermented beverage', *Microbiological Research* (2022); 260

134 **so are beneficial for the heart**: H. Zhao, 'Effects of plant protein and animal protein on lipid profile, body weight and body mass index on patients with hypercholesterolemia: a systematic review and meta-analysis', *Acta Diabetologica* (2020); 57(10)

136 **contain more diversity of microbes**: Y.-C. Shi, 'The bacterial and fungi microbiota of soy sauce-supplied lactic acid bacteria treated with high-pressure process', *Fermentation* (2022); 8(3)

137 **surprising amount of scientific research**: M. Afzaal, 'Nutritional health perspective of Natto: A critical review', *Biochemical Research International* (2022)

137 **eliminating blood clots and reducing dementia**: D. Li, 'Recent advances in nattokinase-enriched fermented soybean foods: A review', *Foods* (2022); 11(13)

138 **fight off other pathogenic microbes**: F. Saeed F, 'Miso: A traditional nutritious & health-endorsing fermented product', Food Science & Nutrition (2022); 10(12)

139 **compared high and low consumers**: S.-J. Jeong, 'Inverse association of daily fermented soybean paste ("Jang") intake with metabolic syndrome risk, especially body fat and hypertension, in men of a large hospital-based cohort', *Frontiers in Nutrition* (2023); 13(10)

139 **probably holds true for all soy versions**: D. Lo, 'Effect of different fermentation conditions on antioxidant capacity and isoflavones content of soy tempeh', *AIMS Agriculture and Food* (2022); 7(3)

144 **exchange of bacteria and yeast**: A. T. Reese, 'Influences of ingredients and bakers on the bacteria and fungi in sourdough starters and bread', *mSphere* (2020); 5(10)

144 **used microbial genetic analysis**: E. A. Landis, 'The diversity and function of sourdough starter microbiomes', *eLife* (2021); 10

145 **clinical trials of sourdough vs yeast**: L. Ribet, 'Nutritional benefits of sourdoughs: A systematic review', *Advances in Nutrition* (2023); 14(1)

146 **lower sugar spikes with my homemade**: D. P. Johansson, 'Impact of food processing on rye product properties and their in vitro digestion', *European Journal of Nutrition* (2018); 57

146 **those with CD or intolerance**: O. J. Ogilvie, 'A case study of the response of immunogenic gluten peptides to sourdough proteolysis', *Nutrients* (2021); 13

148 **imparting special flavour characteristics**: C. Liu, 'Study on the trend in microbial changes during the fermentation of black tea and its effect on the quality', *Foods* (2023); 12

150 **complexity that comes from fermentation**: M. M. Aboulwafa, 'A comprehensive insight on the health benefits and phytoconstituents of Camellia sinensis and recent approaches for its quality control', *Antioxidants (Basel)* (2019); 8(10)

150 **before enjoying the safer second brew**: Y. Zhang, 'The microbiome and metabolites in fermented Pu-erh tea as revealed by high-throughput sequencing and quantitative multiplex metabolite analysis', *PLoS One* (2016); 11(6)

150 **miraculous-sounding health claims**: T. Koláčková, 'Matcha tea: Analysis of nutritional composition, phenolics and antioxidant activity', *Plant Foods for Human Nutrition* (2020); 75

151 **thought to reduce cancer**: X. Li, 'Association between tea consumption and risk of cancer: A prospective cohort study of 0.5 million Chinese adults', *European Journal of Epidemiology* (2019); 34(8)

151 **no clear benefits in non-Asian**: X. Yang, 'Association between tea consumption and prevention of coronary artery disease: A systematic review and dose-response meta-analysis', *Frontiers in Nutrition* (2022); 9

151 **based mainly on test-tube studies**: C. Musial, 'Beneficial properties of green tea catechins', *International Journal of Molecular Sciences* (2020); 21

151 **potentially harmful in high doses**: EFSA Panel on Food Additives and Nutrient Sources added to Food (ANS), 'Scientific opinion on the safety of green tea catechins', *EFSA Journal* (2018); 16(4)

151 **improve cognition in elderly women**: S. Sokary, 'The therapeutic potential of matcha tea: A critical review on human and animal studies', *Current Research in Food Science* (2022); 23(6)

152 **coffee can prolong your life**: P. Manghi, 'Coffee consumption is associated with intestinal *Lawsonibacter asaccharolyticus* abundance and prevalence across multiple cohorts', *Nature Microbiology* (2024); 9(12)

153 **studies show that the quality of coffee**: R. Cruz-O'Byrne, 'Microbial diversity associated with spontaneous coffee bean fermentation process and specialty coffee production in northern Colombia', *International Journal of Food Microbiology* (2021); 354

153 **as starters during their fermentation**: D. M. de Jesus Cassimiro, 'Wet fermentation of *Coffea canephora* by lactic acid bacteria and yeasts using the self-induced anaerobic fermentation (SIAF) method enhances the coffee quality', *Food Microbiology* (2023); 110

154 **with big variations between beans**: M. Mestanza, 'Changes of polyphenols and antioxidants of arabica coffee varieties during roasting', *Frontiers in Nutrition* (2023); 10

155 **increased markers of biological age**: Y. Wei, 'Instant coffee is negatively associated with telomere length: Finding from observational and mendelian randomization analyses of UK biobank', *Nutrients* (2023); 15(6)

155 **can irritate your bladder**: A. Staack, 'Prospective study on the effects of regular and decaffeinated coffee on urinary symptoms in young and healthy volunteers', *Neurourology Urodynamics* (2017); 36(2)

156 **after bowel surgery**: S. Sinz, 'Gum chewing and coffee consumption but not caffeine intake improve bowel function after gastrointestinal surgery: a systematic review and network meta-analysis', *Journal of Gastrointestinal Surgery* (2023); 27(8)

156 **had mild stimulatory effects**: V. Mehta, 'Effect of caffeine on colonic manometry in children', *Journal of Pediatric Gastroenterology and Nutrition* (2023); 76(1)

156 **part of a recommended treatment plan**: R. D. Heath, 'Coffee: The magical bean for liver diseases', *World Journal of Hepatology* (2017); 9(15)

156 **it produces the helpful short chain fatty acid**: M. Sakamoto, '*Lawsonibacter asaccharolyticus* gen. nov., sp. nov., a butyrate-producing bacterium isolated from human faeces', *International Journal of Systematic and Evolutionary Microbiology* (2018); 68(6)

161 **we all taste wine slightly differently**: C. Muñoz-González, 'Ability of human oral microbiota to produce wine odorant aglycones from odourless grape glycosidic aroma precursors', *Food Chemistry* (2015); 187

162 **before it reaches our guts**: M. Schwartz, 'Impact of oral microbiota on flavor perception: From food processing to in-mouth metabolization', *Foods* (2021); 10(9)

162 **is slightly increased**: A. M. Wood, 'Risk thresholds for alcohol consumption: combined analysis of individual-participant data for 599 912 current drinkers in 83 prospective studies', *The Lancet*, 391(1012)

162 **on your biological age**: A. Topiwala, 'Alcohol consumption and telomere length: Mendelian randomization clarifies alcohol's effects', *Molecular Psychiatry* (2022); 27(10)

162 **no safe minimum level**: GBT 2016 Alcohol Collaborators, 'Alcohol use and burden for 195 countries and territories, 1990–2016: A systematic analysis for the Global Burden of Disease Study 2016', 392(10152)

162 **the colour or types of wine**: M. Lucerón-Lucas-Torres, 'Association between wine consumption with cardiovascular disease and cardiovascular mortality: A systematic review and meta-analysis', *Nutrients* (2023); 15(12)

163 **heart disease in patients with diabetes**: M. Lombardo, 'Health effects of red wine consumption: A narrative review of an issue that still deserves debate', *Nutrients* (2023); 15(8)

163 **antioxidant chemicals in the blood**: E. A. Haas, 'A red wine intervention does not modify plasma trimethylamine N-oxide but is associated with broad shifts in the plasma metabolome and gut microbiota composition', *American Journal of Clinical Nutrition* (2022); 116(6)

163 **reducing dramatically at three glasses**: C. I. Le Roy, 'Red wine consumption associated with increased gut microbiota α-diversity in 3 independent cohorts', *Gastroenterology* (2020); 158(1)

163 **and reduce insulin resistance**: L. Castaldo, 'Red wine consumption and cardiovascular health', *Molecules* (2019); 24(19)

164 **other factors are important**: M. Taborsky, 'Red or white wine consumption effect on atherosclerosis in healthy individuals (In Vino Veritas study)', *Bratislava Medical Journal* (2017); 118(5)

164 **effect of resveratrol on the heart**: X. Zheng, 'Effects of resveratrol supplementation on cardiac remodeling in hypertensive patients: a randomized controlled clinical trial', *Hypertension Research* (2023); 46(6)

165 **mostly showing reductions in blood lipids**: G. C. Batista-Jorge, 'Oral resveratrol supplementation improves Metabolic Syndrome features in obese patients submitted to a lifestyle-changing program', *Life Sciences* (2020); 256

165 **made blood lipids worse**: T. N. Kjær, 'No Beneficial Effects of Resveratrol on the Metabolic Syndrome: A Randomized Placebo-Controlled Clinical Trial', *Journal of Clinical Endocrinology and Metabolism* (2017); 102(5)

165 **beneficial effect on lowering blood lipids**: X. Cao, 'The effect of resveratrol on blood lipid profile: A dose-response meta-analysis of randomized controlled trials', *Nutrients* (2022); 14(18)

165 **and lower cholesterol levels**: L. Briansó-Llort, 'Effect of resveratrol content in red wine on circulating sex hormone-binding globulin: Lessons from a pilot clinical trial', *Molecular Nutrition & Food Research* (2022); 66(16)

165 **can – in contrast – increase body weight**: S. M. Mousavi, 'Resveratrol supplementation significantly influences obesity measures: A systematic review and dose-response meta-analysis of randomized controlled trials', *Obesity Reviews* (2019); 20(3)

165 **1000mg was actually harmful**: H. Cai, 'Cancer chemoprevention: Evidence of a nonlinear dose response for the protective effects of resveratrol in humans and mice', *Science Translational Medicine* (2015); 7(298)

165 **98.4 per cent of resveratrol comes from wine-drinking**: R. Zamora-Ros, 'Concentrations of resveratrol and derivatives in foods and estimation of dietary intake in a Spanish population: European Prospective Investigation into Cancer and Nutrition (EPIC)-Spain cohort', *British Journal of Nutrition* (2008); 100

166 **pinots produce over 13mg**: E. H. Siemann, 'Concentration of the phytoalexin resveratrol in wine', *American Journal Enology and Viticulture* (1992); 43(49)

166 **conversion of resveratrol to less active forms**: M. Naiker, 'Loss of trans-resveratrol during storage and ageing of red wines', *Australian Journal of Grape and Wine Research* (2020); 26

166 **returning to zero after a night on the booze**: J. C. Verster, 'Updating the definition of the alcohol hangover', *Journal of Clinical Medicine* (2020); 9

167 **or fermented ginseng**: J. Fan, 'Fermented ginseng improved alcohol liver injury in association with changes in the gut microbiota of mice', *Food & Function* (2019); 10(9)

167 **counteracting alcoholic liver damage**: T. J. Lim, 'Effects of multi-species probiotic supplementation on alcohol metabolism in rats', *Journal of Microbiology* (2021); 59

167 **because doses were too low**: S. J. Jung, 'Regulation of alcohol and acetaldehyde metabolism by a mixture of *Lactobacillus* and *Bifidobacterium* species in human', *Nutrients* (2021); 13(6)

167 **only before, not after the event**: A. Merlo, 'Proceedings of the 10th Alcohol Hangover Research Group Meeting in Utrecht, The Netherlands', *Proceedings* (2020); 43(4)

168 **and 14 per cent with the low dose**: M. Silva, 'Sulfite concentration and the occurrence of headache in young adults: a prospective study', *European Journal of Clinical Nutrition* (2019); 73

169 **phantom smell in susceptible people**: H. Takeuchi, '2,4,6-trichloroanisole is a potent suppressor of olfactory signal transduction', *Proceedings of the National Academy of Sciences of the United States of America* (2013); 110(40)

170 **the tasters were lost**: F. Brochet, 'Wine descriptive language supports cognitive specificity of chemical senses', *Brain and Language* (2001); 77(2)

172 **showed soil quality was better**: A. Christel, 'Impact of farming systems on soil ecological quality: a meta-analysis', *Environmental Chemistry Letters* (2021); 19

172 **and organic wines over five years**: M. Rienth, 'Effects of biodynamic preparations 500 and 501 on vine and berry physiology, pedology and the soil microbiome', *OENO One* (2023); 57(1)

174 **drinking to excess and alcoholism**: C. C. Chen, 'Interaction between the functional polymorphisms of the alcohol-metabolism genes in protection against alcoholism', *American Journal of Human Genetics* (1999); 65(3)

175 **reports of mice being healthier**: S. Fuloria, 'Synbiotic effects of fermented rice on human health and wellness: a natural beverage that boosts immunity', *Frontiers in Microbiology* (2022); 13

177 **and significantly reduce coughs**: S. Samarghandian, 'Honey and health: A review of recent clinical research', *Pharmacognosy Research* (2017); 9(2)

177 **become enriched in the resulting honey**: L. A. Santorelli, 'Beehives possess their own distinct microbiomes', *Environmental Microbiome* (2023); 18(1)

177 **alongside alcohol-forming yeasts**: S. Bovo, 'Shotgun sequencing of honey DNA can describe honey bee derived environmental signatures and the honey bee hologenome complexity', *Scientific Reports* (2020); 10(1)

180 **could be due to error or bias**: N. Papadimitriou, 'A prospective diet-wide association study for risk of colorectal cancer in EPIC', *Clinical Gastroenterology and Hepatology* (2022); 20(4)

180 **inflammatory and vascular biomarkers**: G. Chiva-Blanch, 'Effects of alcohol and polyphenols from beer on atherosclerotic biomarkers in high cardiovascular risk men: a randomized feeding trial', *Nutrition, Metabolism and Cardiovascular Diseases* (2015); 25(1)

180 **there was no real control group**: C. Marques, 'Impact of beer and nonalcoholic beer consumption on the gut microbiota: A randomized, double-blind, controlled trial', *Journal of Agricultural and Food Chemistry* (2022); 70(41)

180 **beer drinkers after thirty days**: F. Hernández-Quiroz; 'Influence of moderate beer consumption on human gut microbiota and its impact on fasting glucose and β-cell function', *Alcohol* (2020 Jun); 85

181 **either on weight or blood pressure**: B. O. Langley, 'Xanthohumol microbiome and signature in healthy adults (the XMaS Trial): Safety and tolerability results of a Phase I triple-masked, placebo-controlled clinical trial', *Molecular Nutrition & Food Research* (2021); 65(8)

181 **mix of turmeric, pepper and hops**: V. A. Tirado-Kulieva, 'A comprehensive review of the benefits of drinking craft beer: Role of phenolic content in health and possible potential of the alcoholic fraction', *Current Research in Food Science* (2023); 6

181 **the recommended 'healthy' levels**: I. Goñi, 'Dietary fiber in beer: Content, composition, colonic fermentability, and contribution to the diet', *Beer in Health and Disease Prevention*, edited by Victor R. Preedy, Academic Press (2009)

181 **and light beers have the lowest**: J. E. S. J. Reid, 'Non-starch polysaccharides in beer and brewing: A review of their occurrence and significance', *Critical Reviews in Food Science and Nutrition* (2024); 64(3)

181 **previously not been recognised**: C.-A. Zugravu, 'Beer and microbiota: Pathways for a positive and healthy interaction', *Nutrients* (2023); 15

182 **for the perfect 'cleansing' ale**: M. Z. A. Chan, 'Survival of probiotic strain *Lactobacillus paracasei* L26 during co-fermentation with S. cerevisiae for the development of a novel beer beverage', *Food Microbiology* (2019); 82

184 **more diverse in cider than beer**: A. Tyakht, 'Characteristics of bacterial and yeast microbiomes in spontaneous and mixed-fermentation beer and cider', *Food Microbiology* (2021); 94

185 **and calcium as the main minerals**: M. Millet, 'Haze in apple-based beverages: Detailed polyphenol, polysaccharide, protein, and mineral compositions', *Journal of Agricultural and Food Chemistry* (2017); 65(31)

189 **reducing blood lipids and inflammation**: S. Kenig, 'Moderate but not high daily intake of chili pepper sauce improves serum glucose and cholesterol levels', *Journal of Functional Foods* (2018); 44

192 **and boost its pharmacological effects**: A. K. Paul, 'Are fermented foods effective against inflammatory diseases?', *International Journal of Environmental Research and Public Health* (2023); 20

193 **add in some more brewer's yeast**: E. D. Kerr, 'Vegemite beer: yeast extract spreads as nutrient supplements to promote fermentation', *PeerJ* (2016); 4

193 **diet ingredients that increased longevity**: Q. Li, 'Protein-rich yeast extract (®fermgard) has potential antioxidant and anti-aging activities', *Comparative Biochemistry and Physiology Part C: Toxicology & Pharmacology* (2023); 270

193 **brain's response to visual patterns**: A. K. Smith, 'Dietary modulation of cortical excitation and inhibition', *Journal of Psychopharmacology* (2017); 31(5)

## PART FOUR

198 **using precision fermentation**: Y. Wang, 'Microbial cell factories for green production of vitamins', *Frontiers in Bioengineering and Biotechnology* (2021); 9

199 **can also produce valuable pharmaceuticals**: C. A. Voigt, 'Synthetic biology 2020–2030: six commercially-available products that are changing our world', *Nature Communications* (2020); 11

199 **halve our agricultural carbon emissions**: F. Humpenöder, 'Projected environmental benefits of replacing beef with microbial protein', *Nature* (2022); 605

202 **more readily than meat alternatives**: P. Slade, 'Killing the sacred dairy cow? Consumer preferences for plant-based milk alternatives', *Agribusiness* (2024); 40

203 **This yeast alone has over 150 specific genes**: S. T. Coradetti, 'Functional genomics of lipid metabolism in the oleaginous yeast *Rhodosporidium toruloides*', *eLife* (2018); 7

204 **seaweed to make biodegradable plastics**: P. M. Noronha, 'Feeding bacteria seaweed to make compostable plastic', Nature.com (2023)

204 **even make electricity in the process**: S. Kalathil, 'Microbial fermentation of polyethylene terephthalate (PET) plastic waste for the production of chemicals or electricity', *Angewandte Chemie* (2022); 134(45)

205 **acting as perfect prevention**: S. M. Brooks, 'Applications, challenges, and needs for employing synthetic biology beyond the lab', *Nature Communications* (2021); 12

205 **to destroy them and eliminate the cancer**: S. Liang, 'Recent advances in bacteria-mediated cancer therapy', *Frontiers in Bioengineering and Biotechnology* (2022); 10

205 **thereby offering a personalised approach**: M. T. Khan, 'Synergy and oxygen adaptation for development of next-generation probiotics', *Nature* (2023); 620

205 **study to reduce blood sugar levels**: P. W. Gilijamse, 'Treatment with *Anaerobutyricum soehngenii*: a pilot study of safety and dose-response effects on glucose metabolism in human subjects with metabolic syndrome', *NPJ Biofilms and Microbiomes* (2020); 6(1)

207 **into a probiotic tomato juice**: P. Cichońska, 'The survival of psychobiotics in fermented food and the gastrointestinal tract: a review', *Microorganisms* (2023); 11(4)

# Acknowledgements

As this book took over six years to write and slowly ferment, many people have been involved in its creation and delivery. As always, my long-term editor at Penguin Random House, Bea Hemming, was crucial in her insight in seeing *Ferment* as a separate project that deserved its own space. She was helped for a year by Jenny Dean, who carefully read the first rough draft and shaped the next. Annie Rigg came on board to expertly test and improve my recipes and fill her kitchen with ferments. Sophie Lambert, my long-standing friend and agent, was there to help and support the project.

Helping me collate all the latest science and shape the book were my friend and colleague, multi-talented nutritionist and scientist Dr Federica Amati, as well as the gifted Dr Lucy McCann, who organised the fermented food sequencing study. Meg Wallace did a great job analysing the ZOE Ferment study so it could be added to this book in time. Nicola Segata's team in Trento did an outstanding job sequencing all the biome studies of prebiotics and all the foods and drinks, as did the King's Twins lab run by Sam Wade in extracting the DNA from the messy samples.

Tim Newman helped me with corrections and the final stages of proofreading, as did Graeme Hall, Jamie Taylor, Clare Sayer and Fiona Brown from PRH. Alison Davies and Sarah Bennie also did an amazing PR job and filled up my diary. Chris Bell prepared the index.

I'd like to thank my family and friends for putting up with the experiments I used to test on them and the strange smells and tastes, and with me boring them with my kimchi stories and forced kombucha testing. I also want to thank King's College London for providing me with an academic base for over thirty years and sharing my passion for taking science to the public. ZOE – the science and nutrition company I co-founded in 2017 – and its CEO Jonathan Wolf have always supported my research and writing and share my ambition to transform the health of millions. Finally, here is a list of some of the many other experts whose brains I have picked along the journey, especially those enthusiasts who have dedicated their lives to fermenting. I should also mention that no two suggested recipes from the experts were ever identical or often even close – making fermenting so much fun. It also means that you can change all the proportions yourselves and prove me wrong. Just let me know on my Instagram account @tim-spector

Yuji Akiyama
Francesco Aniscar
Federica Armanini
Fiona Bennie
Leon Bjerregaard
Andrew Bowhay
Andy Braithwaite
Nicky Briggs
Will Bulsiewicz
Jogile Bulavaite
Steve Byrne
Patrice Cani
William Chase
Andrew Clarke
Jelena Deminska
Raj Dey
Rachel de Thample
Julius Fiedler
Hugh Fearnley-Whittingstall
Christopher Gardner
Jack Gilbert
Nicola Hart
James Hoffmann
Justin Horne
Spencer Hyman
Connor Jordan
Sandor Katz
Denise Kelly
Vanessa Kimbell
Andrew Kojima
Uyen Luu
Doug McMaster
Kenji Morimoto
Sam Murphy
Rasmus Munk
Josh Puddle
Rebecca Palmer
James Read
Neil Rankin
Nicola Segata
Justin & Erica Sonnenburg
Akira Suzuki
Daniel Watkins
Harry Watmough
Alan & Mark Wogan

Apologies to my complaining family and other friends and experts I have forgotten to list!

# General index

# Recipe index

## About the author

Tim Spector is Professor of Epidemiology at King's College London. He is the bestselling author of *The Diet Myth*, *Spoon-Fed*, *Food for Life* and *The Food for Life Cookbook*, and scientific co-founder of the nutrition science company ZOE. With a focus on cutting-edge science and honoured with an OBE for his work in fighting Covid-19, Tim stands at the forefront of his field. The original pioneer of microbiome research, he is among the top 100 most cited scientists in the world.